2020 ADVANCED AUTOMOTIVE TECHNOLOGY BUYING GUIDE

Walter Kreucher

2020 Advanced Automotive Technology Buying Guide Copyright © 2020 by Walter Kreucher All Rights Reserved.

All rights reserved. No part of this book may be reproduced in any form or by any electronic or mechanical means including information storage and retrieval systems, without permission in writing from the author. The only exception is by a reviewer, who may quote short excerpts in a review.

Cover designed by Walter Kreucher
Cover Image is the Mach E

This book is a work of non-fiction.

Walter Kreucher
Visit my website at walt.kreucher.net

Printed in the United States of America

First Printing: February 2020
Dandelion Man Press

ISBN-9798600257221

ELECTRIC VEHICLES

The 2020 Model Year[1] EPA/NHTSA Fuel Economy Guide contains information on twenty-six battery electric vehicles. All of these vehicles met their 2020 model year fuel economy targets and their 2025 model year fuel economy targets.

None of the battery electric vehicles saved the customer money using the methodology employed by the National Highway Traffic Safety Administration and EPA for calculating the cost and benefits of technology.

The average cost premium (including the social benefits) was $20,500.

The EPA reports their estimate for the driving range for electric vehicles. These estimates should be viewed with caution as they represent the maximum driving range under ideal condition of 72°F with all accessories including the heater and the air conditioning system in the off position.

AAA published a study[2] where they tested six electric vehicles at 20° with the heater turned on. The range decreased by an average of 41% (48% compared to the range listed in the EPA mileage guide). AAA also tested the vehicles at 95° with the air conditioning turned on. The range decreased by an average of 17%.

The costs to operate an electric vehicle in this study rely on EPA testing conducted at 72° with the heater, air conditioner and all accessories turned off. The AAA study demonstrates that with the heater or air conditioner turned on the cost of driving an electric vehicle can more than double.

The cost of a replacement battery is also not included in this assessment. It has been estimated[3] that Tesla's cost for a 75-kW battery pack is $16,000. Anyone who owns a cell phone or a laptop computer knows that batteries deteriorate with time and require replacement.

---

[1] As of January 3, 2020
[2] AAA Electric Vehicle Range Testing, February 2019,
https://www.aaa.com/AAA/common/AAR/files/AAA-Electric-Vehicle-Range-Testing-Report.pdf
[3] Sue Babinec, Senior Commercialization Advisor at the Advanced Research Projects Agency-Energy (ARPA-E) at the April 30, 2019 Future Mobility Workshop: Economic and Social Impacts hosted by the University of Michigan's Energy Institute

Also not included are the registration surcharges of up to $200 for electrified vehicles that a growing number of states are enforcing. Twenty-one states[4], including California, charge an annual fee for electric vehicles, most charge between $100 and $150.

---

[4]https://www.bridgemi.com/michigan-environment-watch/electric-car-fees-michigan-would-soar-under-whitmers-roads-plan

Chevrolet Bolt

MSRP: $ 36,620
Gasoline Counterpart: Sonic Hatchback

Lifetime Economic Costs	
Technology Cost	$ 18,600
Net Fuel Cost (Savings)	$ (5,262)
Net Taxes & Fees	$ 1,016
Net Financing	$ 2,850
Net Insurance	$ 3,577
Relative Value Loss	$ 361
Net Cost (Savings)	$ 21,141

Social Benefits	
Refueling Time Savings	$ 305
Energy Security	$ 430
Social Cost of Carbon Emissions	$ 2,041
Crashes, Fatalities, Congestion, Noise	$ (44)
Lost Fuel Tax Revenue	$ (1,614)
Net Cost (Savings) Including Social Benefits	$ 20,024

Meets 2020 Target	✓
Meets Original 2025 Target	✓
Saves Money (including Social Benefits)	✗
Breakeven cost of Gasoline (per Gallon)	$ 6.24

All Electric Range (Miles) 72° HVAC OFF per EPA	259
All Electric Driving Range (Miles) Winter (AAA Study)	123
All Electric Driving Range (Miles) Summer (AAA Study)	143

Assumes Diesel and Electricity Prices do not change in relation to gasoline

Electric Vehicle

Jaguar I-Pace

MSRP: $ 69,850
Gasoline Counterpart: E-Pace

Lifetime Economic Costs	
Technology Cost	$ 29,900
Net Fuel Cost (Savings)	$ (5,112)
Net Taxes & Fees	$ 1,633
Net Financing	$ 4,581
Net Insurance	$ 5,750
Relative Value Loss	$ 580
Net Cost (Savings)	$ 37,331

Social Benefits	
Refueling Time Savings	$ 191
Energy Security	$ 270
Social Cost of Carbon Emissions	$ 2,607
Crashes, Fatalities, Congestion, Noise	$ (43)
Lost Fuel Tax Revenue	$ (2,027)
Net Cost (Savings) Including Social Benefits	$ 36,334

Meets 2020 Target	✓
Meets Original 2025 Target	✓
Saves Money (including Social Benefits)	✗
Breakeven cost of Gasoline (per Gallon)	$ 7.49

All Electric Range (Miles) 72° HVAC OFF per EPA	234
All Electric Driving Range (Miles) Winter (AAA Study)	111
All Electric Driving Range (Miles) Summer (AAA Study)	129

Assumes Diesel and Electricity Prices do not change in relation to gasoline

Hyundai Kona Electric

MSRP: $ 30,315
Gasoline Counterpart: Kona

Lifetime Economic Costs	
Technology Cost	$ 10,215
Net Fuel Cost (Savings)	$ (6,675)
Net Taxes & Fees	$ 558
Net Financing	$ 1,565
Net Insurance	$ 1,964
Relative Value Loss	$ 198
Net Cost (Savings)	$ 7,826

Social Benefits	
Refueling Time Savings	$ 331
Energy Security	$ 467
Social Cost of Carbon Emissions	$ 2,290
Crashes, Fatalities, Congestion, Noise	$ (56)
Lost Fuel Tax Revenue	$ (1,845)
Net Cost (Savings) Including Social Benefits	$ 6,638

Meets 2020 Target	✓
Meets Original 2025 Target	✓
Saves Money (including Social Benefits)	✗
Breakeven cost of Gasoline (per Gallon)	$ 3.78

All Electric Range (Miles) 72° HVAC OFF per EPA	258
All Electric Driving Range (Miles) Winter (AAA Study)	123
All Electric Driving Range (Miles) Summer (AAA Study)	143

Assumes Diesel and Electricity Prices do not change in relation to gasoline

Hyundai Ioniq Electric

MSRP: $ 30,315
Gasoline Counterpart: Ioniq

Lifetime Economic Costs	
Technology Cost	$ 9,665
Net Fuel Cost (Savings)	$ 302
Net Taxes & Fees	$ 528
Net Financing	$ 1,481
Net Insurance	$ 1,859
Relative Value Loss	$ 188
Net Cost (Savings)	$ 14,022

Social Benefits	
Refueling Time Savings	$ 225
Energy Security	$ 318
Social Cost of Carbon Emissions	$ 753
Crashes, Fatalities, Congestion, Noise	$ 3
Lost Fuel Tax Revenue	$ (484)
Net Cost (Savings) Including Social Benefits	$ 13,207

Meets 2020 Target	✓
Meets Original 2025 Target	✓
Saves Money (including Social Benefits)	✗
Breakeven cost of Gasoline (per Gallon)	$ 7.98

All Electric Range (Miles) 72° HVAC OFF per EPA	258
All Electric Driving Range (Miles) Winter (AAA Study)	123
All Electric Driving Range (Miles) Summer (AAA Study)	143

Assumes Diesel and Electricity Prices do not change in relation to gasoline

KIA 2019 Niro

MSRP: $ 38,500
Gasoline Counterpart: Niro

Lifetime Economic Costs		
Technology Cost	$	15,010
Net Fuel Cost (Savings)	$	(3,608)
Net Taxes & Fees	$	820
Net Financing	$	2,300
Net Insurance	$	2,886
Relative Value Loss	$	291
Net Cost (Savings)	$	17,698

Social Benefits		
Refueling Time Savings	$	186
Energy Security	$	262
Social Cost of Carbon Emissions	$	841
Crashes, Fatalities, Congestion, Noise	$	(30)
Lost Fuel Tax Revenue	$	(533)
Net Cost (Savings) Including Social Benefits	$	16,974

Meets 2020 Target	✓
Meets Original 2025 Target	✓
Saves Money (including Social Benefits)	✗
Breakeven cost of Gasoline (per Gallon)	$ 10.04

All Electric Range (Miles) 72° HVAC OFF per EPA	239
All Electric Driving Range (Miles) Winter (AAA Study)	114
All Electric Driving Range (Miles) Summer (AAA Study)	132

Assumes Diesel and Electricity Prices do not change in relation to gasoline

Electric Vehicle

Kia Soul Electric

MSRP: $ 33,145
Gasoline Counterpart: Soul

Lifetime Economic Costs	
Technology Cost	$ 15,745
Net Fuel Cost (Savings)	$ (5,670)
Net Taxes & Fees	$ 860
Net Financing	$ 2,412
Net Insurance	$ 3,028
Relative Value Loss	$ 305
Net Cost (Savings)	$ 16,680

Social Benefits	
Refueling Time Savings	$ 302
Energy Security	$ 427
Social Cost of Carbon Emissions	$ 2,145
Crashes, Fatalities, Congestion, Noise	$ (48)
Lost Fuel Tax Revenue	$ (1,704)

Net Cost (Savings) Including Social Benefits	$ 15,558

Meets 2020 Target	✓
Meets Original 2025 Target	✓
Saves Money (including Social Benefits)	✗
Breakeven cost of Gasoline (per Gallon)	$ 5.33

All Electric Range (Miles) 72° HVAC OFF per EPA	243
All Electric Driving Range (Miles) Winter (AAA Study)	116
All Electric Driving Range (Miles) Summer (AAA Study)	134

Assumes Diesel and Electricity Prices do not change in relation to gasoline

Electric Vehicles

VW e-Golf

MSRP: $ 31,895
Gasoline Counterpart: Golf

Lifetime Economic Costs	
Technology Cost	$ 10,045
Net Fuel Cost (Savings)	$ (2,206)
Net Taxes & Fees	$ 548
Net Financing	$ 1,539
Net Insurance	$ 1,932
Relative Value Loss	$ 195
Net Cost (Savings)	$ 12,052

Social Benefits	
Refueling Time Savings	$ 252
Energy Security	$ 356
Social Cost of Carbon Emissions	$ 1,462
Crashes, Fatalities, Congestion, Noise	$ (19)
Lost Fuel Tax Revenue	$ (1,086)
Net Cost (Savings) Including Social Benefits	$ 11,088

Meets 2020 Target	✓
Meets Original 2025 Target	✓
Saves Money (including Social Benefits)	✗
Breakeven cost of Gasoline (per Gallon)	$ 5.29

All Electric Range (Miles) 72° HVAC OFF per EPA	123
All Electric Driving Range (Miles) Winter (AAA Study)	59
All Electric Driving Range (Miles) Summer (AAA Study)	68

Assumes Diesel and Electricity Prices do not change in relation to gasoline

PORSCHE TAYCAN TURBO

MSRP: $ 150,900
Gasoline Counterpart: Panamera

Lifetime Economic Costs		
Technology Cost	$	63,700
Net Fuel Cost (Savings)	$	2,350
Net Taxes & Fees	$	3,478
Net Financing	$	9,759
Net Insurance	$	12,250
Relative Value Loss	$	1,236
Net Cost (Savings)	$	92,772

Social Benefits		
Refueling Time Savings	$	150
Energy Security	$	211
Social Cost of Carbon Emissions	$	1,950
Crashes, Fatalities, Congestion, Noise	$	20
Lost Fuel Tax Revenue	$	(1,416)
Net Cost (Savings) Including Social Benefits		$ 91,858

Meets 2020 Target	✓
Meets Original 2025 Target	✓
Saves Money (including Social Benefits)	✗
Breakeven cost of Gasoline (per Gallon)	$ 17.63

All Electric Range (Miles) 72° HVAC OFF per EPA	201
All Electric Driving Range (Miles) Winter (AAA Study)	96
All Electric Driving Range (Miles) Summer (AAA Study)	111

Assumes Diesel and Electricity Prices do not change in relation to gasoline

PORSCHE TAYCAN TURBO S

MSRP: $ 185,000
Gasoline Counterpart: Panamera Turbo

Lifetime Economic Costs		
Technology Cost	$	32,000
Net Fuel Cost (Savings)	$	2,663
Net Taxes & Fees	$	1,747
Net Financing	$	4,902
Net Insurance	$	6,154
Relative Value Loss	$	621
Net Cost (Savings)	$	48,087

Social Benefits		
Refueling Time Savings	$	144
Energy Security	$	203
Social Cost of Carbon Emissions	$	1,932
Crashes, Fatalities, Congestion, Noise	$	22
Lost Fuel Tax Revenue	$	(1,394)
Net Cost (Savings) Including Social Benefits		$ 47,179

Meets 2020 Target	✓
Meets Original 2025 Target	✓
Saves Money (including Social Benefits)	✗
Breakeven cost of Gasoline (per Gallon)	$ 10.40

All Electric Range (Miles) 72° HVAC OFF per EPA	192
All Electric Driving Range (Miles) Winter (AAA Study)	91
All Electric Driving Range (Miles) Summer (AAA Study)	106

Assumes Diesel and Electricity Prices do not change in relation to gasoline

Electric Vehicle

BYD e6

MSRP: $ 52,000
Gasoline Counterpart: Average Comparable Non-BEV

Lifetime Economic Costs		
Technology Cost	$	18,390
Net Fuel Cost (Savings)	$	882
Net Taxes & Fees	$	1,004
Net Financing	$	2,817
Net Insurance	$	3,536
Relative Value Loss	$	357
Net Cost (Savings)	$	26,987

Social Benefits		
Refueling Time Savings	$	134
Energy Security	$	190
Social Cost of Carbon Emissions	$	1,444
Crashes, Fatalities, Congestion, Noise	$	7
Lost Fuel Tax Revenue	$	(975)
Net Cost (Savings) Including Social Benefits	$	26,186

Meets 2020 Target	✓
Meets Original 2025 Target	✓
Saves Money (including Social Benefits)	✗
Breakeven cost of Gasoline (per Gallon)	$ 8.08

All Electric Range (Miles) 72° HVAC OFF per EPA	187
All Electric Driving Range (Miles) Winter (AAA Study)	89
All Electric Driving Range (Miles) Summer (AAA Study)	103

Assumes Diesel and Electricity Prices do not change in relation to gasoline

Electric Vehicles

Tesla Model 3 Long Range

MSRP: $ 48,990
Gasoline Counterpart: Average Comparable Non-BEV

Lifetime Economic Costs		
Technology Cost	$	18,390
Net Fuel Cost (Savings)	$	(4,363)
Net Taxes & Fees	$	1,004
Net Financing	$	2,817
Net Insurance	$	3,536
Relative Value Loss	$	357
Net Cost (Savings)	$	21,742

Social Benefits		
Refueling Time Savings	$	330
Energy Security	$	466
Social Cost of Carbon Emissions	$	1,747
Crashes, Fatalities, Congestion, Noise	$	(37)
Lost Fuel Tax Revenue	$	(1,369)
Net Cost (Savings) Including Social Benefits		$ 20,605

Meets 2020 Target	✓
Meets Original 2025 Target	✓
Saves Money (including Social Benefits)	✗
Breakeven cost of Gasoline (per Gallon)	$ 6.95

All Electric Range (Miles) 72° HVAC OFF per EPA	330
All Electric Driving Range (Miles) Winter (AAA Study)	157
All Electric Driving Range (Miles) Summer (AAA Study)	182

Assumes Diesel and Electricity Prices do not change in relation to gasoline

Electric Vehicle

Tesla Model 3 Long Range AWD

MSRP: $ 48,990
Gasoline Counterpart: Average Comparable Non-BEV

Lifetime Economic Costs		
Technology Cost	$	18,390
Net Fuel Cost (Savings)	$	(3,805)
Net Taxes & Fees	$	1,004
Net Financing	$	2,817
Net Insurance	$	3,536
Relative Value Loss	$	357
Net Cost (Savings)	$	22,299

Social Benefits		
Refueling Time Savings	$	296
Energy Security	$	417
Social Cost of Carbon Emissions	$	1,714
Crashes, Fatalities, Congestion, Noise	$	(32)
Lost Fuel Tax Revenue	$	(1,326)
Net Cost (Savings) Including Social Benefits		$ 21,230

Meets 2020 Target		✓
Meets Original 2025 Target		✓
Saves Money (including Social Benefits)		✗
Breakeven cost of Gasoline (per Gallon)	$	7.07

All Electric Range (Miles) 72° HVAC OFF per EPA		322
All Electric Driving Range (Miles) Winter (AAA Study)		153
All Electric Driving Range (Miles) Summer (AAA Study)		178

Assumes Diesel and Electricity Prices do not change in relation to gasoline

Tesla Model 3 Mid Range

MSRP: $ 48,990
Gasoline Counterpart: Average Comparable Non-BEV

Lifetime Economic Costs	
Technology Cost	$ 18,390
Net Fuel Cost (Savings)	$ (3,957)
Net Taxes & Fees	$ 1,004
Net Financing	$ 2,817
Net Insurance	$ 3,536
Relative Value Loss	$ 357
Net Cost (Savings)	$ 22,147

Social Benefits	
Refueling Time Savings	$ 304
Energy Security	$ 429
Social Cost of Carbon Emissions	$ 1,723
Crashes, Fatalities, Congestion, Noise	$ (33)
Lost Fuel Tax Revenue	$ (1,338)
Net Cost (Savings) Including Social Benefits	$ 21,061

Meets 2020 Target	✓
Meets Original 2025 Target	✓
Saves Money (including Social Benefits)	✗
Breakeven cost of Gasoline (per Gallon)	$ 7.03

All Electric Range (Miles) 72° HVAC OFF per EPA	264
All Electric Driving Range (Miles) Winter (AAA Study)	126
All Electric Driving Range (Miles) Summer (AAA Study)	146

Assumes Diesel and Electricity Prices do not change in relation to gasoline

Electric Vehicle

Tesla Model 3 Standard Range

MSRP: $ 40,230
Gasoline Counterpart: Average Comparable Non-BEV

Lifetime Economic Costs		
Technology Cost	$	18,390
Net Fuel Cost (Savings)	$	(4,411)
Net Taxes & Fees	$	1,004
Net Financing	$	2,817
Net Insurance	$	3,536
Relative Value Loss	$	357
Net Cost (Savings)	$	21,694

Social Benefits		
Refueling Time Savings	$	333
Energy Security	$	470
Social Cost of Carbon Emissions	$	1,749
Crashes, Fatalities, Congestion, Noise	$	(37)
Lost Fuel Tax Revenue	$	(1,373)
Net Cost (Savings) Including Social Benefits	$	20,551

Meets 2020 Target	✓	
Meets Original 2025 Target	✓	
Saves Money (including Social Benefits)	✗	
Breakeven cost of Gasoline (per Gallon)	$	6.94

All Electric Range (Miles) 72° HVAC OFF per EPA		220
All Electric Driving Range (Miles) Winter (AAA Study)		105
All Electric Driving Range (Miles) Summer (AAA Study)		122

Tesla Model 3 Standard Range Plus

MSRP: $ 40,230
Gasoline Counterpart: Average Comparable Non-BEV

Lifetime Economic Costs	
Technology Cost	$ 18,390
Net Fuel Cost (Savings)	$ (4,796)
Net Taxes & Fees	$ 1,004
Net Financing	$ 2,817
Net Insurance	$ 3,536
Relative Value Loss	$ 357
Net Cost (Savings)	$ 21,309

Social Benefits	
Refueling Time Savings	$ 361
Energy Security	$ 510
Social Cost of Carbon Emissions	$ 1,772
Crashes, Fatalities, Congestion, Noise	$ (40)
Lost Fuel Tax Revenue	$ (1,402)
Net Cost (Savings) Including Social Benefits	$ 20,109

Meets 2020 Target	✓
Meets Original 2025 Target	✓
Saves Money (including Social Benefits)	✗
Breakeven cost of Gasoline (per Gallon)	$ 6.85

All Electric Range (Miles) 72° HVAC OFF per EPA	250
All Electric Driving Range (Miles) Winter (AAA Study)	119
All Electric Driving Range (Miles) Summer (AAA Study)	138

Assumes Diesel and Electricity Prices do not change in relation to gasoline

Electric Vehicle

Tesla Model 3 Long Range Perf. AWD (18")

MSRP: $ 56,990
Gasoline Counterpart: Average Comparable Non-BEV

Lifetime Economic Costs	
Technology Cost	$ 18,390
Net Fuel Cost (Savings)	$ (4,023)
Net Taxes & Fees	$ 1,004
Net Financing	$ 2,817
Net Insurance	$ 3,536
Relative Value Loss	$ 357
Net Cost (Savings)	$ 22,081

Social Benefits	
Refueling Time Savings	$ 308
Energy Security	$ 435
Social Cost of Carbon Emissions	$ 1,727
Crashes, Fatalities, Congestion, Noise	$ (34)
Lost Fuel Tax Revenue	$ (1,343)
Net Cost (Savings) Including Social Benefits	$ 20,987

Meets 2020 Target	✓
Meets Original 2025 Target	✓
Saves Money (including Social Benefits)	✗
Breakeven cost of Gasoline (per Gallon)	$ 7.02

All Electric Range (Miles) 72° HVAC OFF per EPA	322
All Electric Driving Range (Miles) Winter (AAA Study)	153
All Electric Driving Range (Miles) Summer (AAA Study)	178

Assumes Diesel and Electricity Prices do not change in relation to gasoline

Electric Vehicles

Tesla Model 3 Long Range Perf. AWD (19")

MSRP: $ 56,990
Gasoline Counterpart: Average Comparable Non-BEV

Lifetime Economic Costs	
Technology Cost	$ 18,390
Net Fuel Cost (Savings)	$ (3,239)
Net Taxes & Fees	$ 1,004
Net Financing	$ 2,817
Net Insurance	$ 3,536
Relative Value Loss	$ 357
Net Cost (Savings)	$ 22,866

Social Benefits	
Refueling Time Savings	$ 266
Energy Security	$ 375
Social Cost of Carbon Emissions	$ 1,682
Crashes, Fatalities, Congestion, Noise	$ (27)
Lost Fuel Tax Revenue	$ (1,282)
Net Cost (Savings) Including Social Benefits	$ 21,853

Meets 2020 Target	✓
Meets Original 2025 Target	✓
Saves Money (including Social Benefits)	✗
Breakeven cost of Gasoline (per Gallon)	$ 7.19

All Electric Range (Miles) 72° HVAC OFF per EPA	304
All Electric Driving Range (Miles) Winter (AAA Study)	145
All Electric Driving Range (Miles) Summer (AAA Study)	168

Assumes Diesel and Electricity Prices do not change in relation to gasoline

Tesla Model 3 Long Range Perf. AWD (20")

MSRP: $ 56,990
Gasoline Counterpart: Average Comparable Non-BEV

Lifetime Economic Costs		
Technology Cost	$	18,390
Net Fuel Cost (Savings)	$	(3,084)
Net Taxes & Fees	$	1,004
Net Financing	$	2,817
Net Insurance	$	3,536
Relative Value Loss	$	357
Net Cost (Savings)	$	23,021

Social Benefits		
Refueling Time Savings	$	258
Energy Security	$	364
Social Cost of Carbon Emissions	$	1,673
Crashes, Fatalities, Congestion, Noise	$	(26)
Lost Fuel Tax Revenue	$	(1,270)
Net Cost (Savings) Including Social Benefits		$ 22,023

Meets 2020 Target	✓
Meets Original 2025 Target	✓
Saves Money (including Social Benefits)	✗
Breakeven cost of Gasoline (per Gallon)	$ 7.22

All Electric Range (Miles) 72° HVAC OFF per EPA	299
All Electric Driving Range (Miles) Winter (AAA Study)	142
All Electric Driving Range (Miles) Summer (AAA Study)	165

Assumes Diesel and Electricity Prices do not change in relation to gasoline

Tesla Model S Long Range

MSRP: $ 79,990
Gasoline Counterpart: Average Comparable Non-BEV

Lifetime Economic Costs	
Technology Cost	$ 18,390
Net Fuel Cost (Savings)	$ (2,343)
Net Taxes & Fees	$ 1,004
Net Financing	$ 2,817
Net Insurance	$ 3,536
Relative Value Loss	$ 357
Net Cost (Savings)	$ 23,762

Social Benefits	
Refueling Time Savings	$ 242
Energy Security	$ 342
Social Cost of Carbon Emissions	$ 1,652
Crashes, Fatalities, Congestion, Noise	$ (20)
Lost Fuel Tax Revenue	$ (1,244)
Net Cost (Savings) Including Social Benefits	$ 22,789

Meets 2020 Target	✓
Meets Original 2025 Target	✓
Saves Money (including Social Benefits)	✗
Breakeven cost of Gasoline (per Gallon)	$ 7.38

All Electric Range (Miles) 72° HVAC OFF per EPA	373
All Electric Driving Range (Miles) Winter (AAA Study)	177
All Electric Driving Range (Miles) Summer (AAA Study)	206

Assumes Diesel and Electricity Prices do not change in relation to gasoline

Tesla Model S Performance (19" Wheels)

MSRP: $ 79,990
Gasoline Counterpart: Average Comparable Non-BEV

Lifetime Economic Costs	
Technology Cost	$ 18,390
Net Fuel Cost (Savings)	$ (2,062)
Net Taxes & Fees	$ 1,004
Net Financing	$ 2,817
Net Insurance	$ 3,536
Relative Value Loss	$ 357
Net Cost (Savings)	$ 24,043

Social Benefits	
Refueling Time Savings	$ 216
Energy Security	$ 304
Social Cost of Carbon Emissions	$ 1,614
Crashes, Fatalities, Congestion, Noise	$ (17)
Lost Fuel Tax Revenue	$ (1,193)
Net Cost (Savings) Including Social Benefits	$ 23,120

Meets 2020 Target	✓
Meets Original 2025 Target	✓
Saves Money (including Social Benefits)	✗
Breakeven cost of Gasoline (per Gallon)	$ 7.44

All Electric Range (Miles) 72° HVAC OFF per EPA	348
All Electric Driving Range (Miles) Winter (AAA Study)	166
All Electric Driving Range (Miles) Summer (AAA Study)	192

Assumes Diesel and Electricity Prices do not change in relation to gasoline

Tesla Model S Performance (21" Wheels)

MSRP: $ 99,990
Gasoline Counterpart: Average Comparable Non-BEV

Lifetime Economic Costs	
Technology Cost	$ 18,390
Net Fuel Cost (Savings)	$ (3,084)
Net Taxes & Fees	$ 1,004
Net Financing	$ 2,817
Net Insurance	$ 3,536
Relative Value Loss	$ 357
Net Cost (Savings)	$ 23,021

Social Benefits	
Refueling Time Savings	$ 258
Energy Security	$ 364
Social Cost of Carbon Emissions	$ 1,673
Crashes, Fatalities, Congestion, Noise	$ (26)
Lost Fuel Tax Revenue	$ (1,270)
Net Cost (Savings) Including Social Benefits	$ 22,023

Meets 2020 Target	✓
Meets Original 2025 Target	✓
Saves Money (including Social Benefits)	✗
Breakeven cost of Gasoline (per Gallon)	$ 7.22

All Electric Range (Miles) 72° HVAC OFF per EPA	299
All Electric Driving Range (Miles) Winter (AAA Study)	142
All Electric Driving Range (Miles) Summer (AAA Study)	165

Assumes Diesel and Electricity Prices do not change in relation to gasoline

Electric Vehicle

Tesla Model S Standard Range

MSRP: $ 79,990
Gasoline Counterpart: Average Comparable Non-BEV

Lifetime Economic Costs		
Technology Cost	$	18,390
Net Fuel Cost (Savings)	$	(2,584)
Net Taxes & Fees	$	1,004
Net Financing	$	2,817
Net Insurance	$	3,536
Relative Value Loss	$	357
Net Cost (Savings)	$	23,521

Social Benefits		
Refueling Time Savings	$	236
Energy Security	$	333
Social Cost of Carbon Emissions	$	1,644
Crashes, Fatalities, Congestion, Noise	$	(22)
Lost Fuel Tax Revenue	$	(1,232)
Net Cost (Savings) Including Social Benefits		$ 22,562

Meets 2020 Target	✓
Meets Original 2025 Target	✓
Saves Money (including Social Benefits)	✗
Breakeven cost of Gasoline (per Gallon)	$ 7.33

All Electric Range (Miles) 72° HVAC OFF per EPA	287
All Electric Driving Range (Miles) Winter (AAA Study)	137
All Electric Driving Range (Miles) Summer (AAA Study)	159

Assumes Diesel and Electricity Prices do not change in relation to gasoline

Tesla Model X Long Range

MSRP: $ 84,990
Gasoline Counterpart: Average Comparable Non-BEV

Lifetime Economic Costs	
Technology Cost	$ 18,390
Net Fuel Cost (Savings)	$ (1,689)
Net Taxes & Fees	$ 1,004
Net Financing	$ 2,817
Net Insurance	$ 3,536
Relative Value Loss	$ 357
Net Cost (Savings)	$ 24,415

Social Benefits	
Refueling Time Savings	$ 195
Energy Security	$ 274
Social Cost of Carbon Emissions	$ 1,579
Crashes, Fatalities, Congestion, Noise	$ (14)
Lost Fuel Tax Revenue	$ (1,147)
Net Cost (Savings) Including Social Benefits	$ 23,529

Meets 2020 Target	✓
Meets Original 2025 Target	✓
Saves Money (including Social Benefits)	✗
Breakeven cost of Gasoline (per Gallon)	$ 7.11

All Electric Range (Miles) 72° HVAC OFF per EPA	328
All Electric Driving Range (Miles) Winter (AAA Study)	156
All Electric Driving Range (Miles) Summer (AAA Study)	181

Assumes Diesel and Electricity Prices do not change in relation to gasoline

Electric Vehicle

Tesla Model X Performance (20" Wheels)

MSRP: $ 104,990
Gasoline Counterpart: Average Comparable Non-BEV

Lifetime Economic Costs		
Technology Cost	$	18,390
Net Fuel Cost (Savings)	$	(958)
Net Taxes & Fees	$	1,004
Net Financing	$	2,817
Net Insurance	$	3,536
Relative Value Loss	$	357
Net Cost (Savings)	$	25,147

Social Benefits		
Refueling Time Savings	$	174
Energy Security	$	246
Social Cost of Carbon Emissions	$	1,540
Crashes, Fatalities, Congestion, Noise	$	(8)
Lost Fuel Tax Revenue	$	(1,097)
Net Cost (Savings) Including Social Benefits		$ 24,291

Meets 2020 Target	✓
Meets Original 2025 Target	✓
Saves Money (including Social Benefits)	✗
Breakeven cost of Gasoline (per Gallon)	$ 7.25

All Electric Range (Miles) 72° HVAC OFF per EPA	305
All Electric Driving Range (Miles) Winter (AAA Study)	145
All Electric Driving Range (Miles) Summer (AAA Study)	168

Assumes Diesel and Electricity Prices do not change in relation to gasoline

Electric Vehicles

Tesla Model X Performance (22" Wheels)

MSRP: $ 104,990
Gasoline Counterpart: Average Comparable Non-BEV

Lifetime Economic Costs	
Technology Cost	$ 18,390
Net Fuel Cost (Savings)	$ 600
Net Taxes & Fees	$ 1,004
Net Financing	$ 2,817
Net Insurance	$ 3,536
Relative Value Loss	$ 357
Net Cost (Savings)	$ 26,704

Social Benefits	
Refueling Time Savings	$ 140
Energy Security	$ 197
Social Cost of Carbon Emissions	$ 1,459
Crashes, Fatalities, Congestion, Noise	$ 5
Lost Fuel Tax Revenue	$ (993)
Net Cost (Savings) Including Social Benefits	$ 25,897

Meets 2020 Target	✓
Meets Original 2025 Target	✓
Saves Money (including Social Benefits)	✗
Breakeven cost of Gasoline (per Gallon)	$ 7.56

All Electric Range (Miles) 72° HVAC OFF per EPA	272
All Electric Driving Range (Miles) Winter (AAA Study)	129
All Electric Driving Range (Miles) Summer (AAA Study)	150

Assumes Diesel and Electricity Prices do not change in relation to gasoline

Tesla Model X Standard Range

MSRP: $ 84,990
Gasoline Counterpart: Average Comparable Non-BEV

Lifetime Economic Costs	
Technology Cost	$ 18,390
Net Fuel Cost (Savings)	$ (2,213)
Net Taxes & Fees	$ 1,004
Net Financing	$ 2,817
Net Insurance	$ 3,536
Relative Value Loss	$ 357
Net Cost (Savings)	$ 23,891

Social Benefits	
Refueling Time Savings	$ 211
Energy Security	$ 297
Social Cost of Carbon Emissions	$ 1,606
Crashes, Fatalities, Congestion, Noise	$ (19)
Lost Fuel Tax Revenue	$ (1,183)
Net Cost (Savings) Including Social Benefits	$ 22,979

Meets 2020 Target	✓
Meets Original 2025 Target	✓
Saves Money (including Social Benefits)	✗
Breakeven cost of Gasoline (per Gallon)	$ 7.00

All Electric Range (Miles) 72° HVAC OFF per EPA	258
All Electric Driving Range (Miles) Winter (AAA Study)	123
All Electric Driving Range (Miles) Summer (AAA Study)	143

Assumes Diesel and Electricity Prices do not change in relation to gasoline

PLUG-IN HYBRID ELECTRIC VEHICLES

The 2020 Model Year EPA/NHTSA Fuel Economy Guide contains information on thirty plug-in hybrid electric vehicles. **Twenty-six of these vehicles met their 2020 model year fuel economy targets** and sixty-three percent (nineteen) met their 2025 model year fuel economy targets.

None of the vehicles saved the customer money using the methodology employed by the National Highway Traffic Safety Administration and EPA for calculating the cost and benefits of technology.

The average cost premium for a PHEV was $21,600.

The EPA estimated all electric driving range has been added to the report. This range should be viewed with caution as it represents the range when the vehicle is operated at 72° with the heater and the air conditioner and all accessories turned off. Many of the plug-in hybrid electric vehicles require the engine to be turned on to operate the cabin climate controls. This would bring the all-electric range to zero when the heater or air conditioner is turned on.

BMW 530e

MSRP: $ 53,900
Gasoline Counterpart: 530i

Lifetime Economic Costs		
Technology Cost	$	-
Net Fuel Cost (Savings)	$	696
Net Taxes & Fees	$	-
Net Financing	$	-
Net Insurance	$	-
Relative Value Loss	$	-
Net Cost (Savings)	$	696

Social Benefits		
Refueling Time Savings	$	(18)
Energy Security	$	(25)
Social Cost of Carbon Emissions	$	312
Crashes, Fatalities, Congestion, Noise	$	6
Lost Fuel Tax Revenue	$	(713)
Net Cost (Savings) Including Social Benefits	$	1,134

Meets 2020 Target	✓
Meets Original 2025 Target	✓
Saves Money (including Social Benefits)	✗
Breakeven cost of Gasoline (per Gallon)	NA

All Electric Range (Miles) 72° HVAC OFF per EPA	21
All Electric Driving Range (Miles) Winter (AAA Study)	10
All Electric Driving Range (Miles) Summer (AAA Study)	12

Assumes Diesel and Electricity Prices do not change in relation to gasoline

BMW 530e xDrive

MSRP: $ 56,200
Gasoline Counterpart: 530i xDrive

Lifetime Economic Costs		
Technology Cost	$	-
Net Fuel Cost (Savings)	$	811
Net Taxes & Fees	$	-
Net Financing	$	-
Net Insurance	$	-
Relative Value Loss	$	-
Net Cost (Savings)	$	811

Social Benefits		
Refueling Time Savings	$	(17)
Energy Security	$	(23)
Social Cost of Carbon Emissions	$	317
Crashes, Fatalities, Congestion, Noise	$	7
Lost Fuel Tax Revenue	$	(717)
Net Cost (Savings) Including Social Benefits	$	1,244

Meets 2020 Target	✓
Meets Original 2025 Target	✗
Saves Money (including Social Benefits)	✗
Breakeven cost of Gasoline (per Gallon)	NA

All Electric Range (Miles) 72° HVAC OFF per EPA	19
All Electric Driving Range (Miles) Winter (AAA Study)	9
All Electric Driving Range (Miles) Summer (AAA Study)	10

Assumes Diesel and Electricity Prices do not change in relation to gasoline

BMW 745e xDrive

MSRP: $ 95,550
Gasoline Counterpart: 740i xDrive

Lifetime Economic Costs		
Technology Cost	$	6,100
Net Fuel Cost (Savings)	$	559
Net Taxes & Fees	$	333
Net Financing	$	935
Net Insurance	$	1,173
Relative Value Loss	$	-
Net Cost (Savings)	$	9,099

Social Benefits		
Refueling Time Savings	$	(11)
Energy Security	$	(15)
Social Cost of Carbon Emissions	$	403
Crashes, Fatalities, Congestion, Noise	$	5
Lost Fuel Tax Revenue	$	(842)

Net Cost (Savings) Including Social Benefits	$	9,560

Meets 2020 Target	✓
Meets Original 2025 Target	✗
Saves Money (including Social Benefits)	✗
Breakeven cost of Gasoline (per Gallon)	NA

All Electric Range (Miles) 72° HVAC OFF per EPA	16
All Electric Driving Range (Miles) Winter (AAA Study)	8
All Electric Driving Range (Miles) Summer (AAA Study)	9

Assumes Diesel and Electricity Prices do not change in relation to gasoline

BMW I8 coupe

MSRP: $ 147,500
Gasoline Counterpart: 8 Series Coupe

Lifetime Economic Costs	
Technology Cost	$ 59,600
Net Fuel Cost (Savings)	$ (79)
Net Taxes & Fees	$ 3,254
Net Financing	$ 9,131
Net Insurance	$ 11,461
Relative Value Loss	$ -
Net Cost (Savings)	$ 83,367

Social Benefits	
Refueling Time Savings	$ (9)
Energy Security	$ (13)
Social Cost of Carbon Emissions	$ 477
Crashes, Fatalities, Congestion, Noise	$ (1)
Lost Fuel Tax Revenue	$ (822)
Net Cost (Savings) Including Social Benefits	$ 83,735

Meets 2020 Target	✓
Meets Original 2025 Target	✗
Saves Money (including Social Benefits)	✗
Breakeven cost of Gasoline (per Gallon)	NA

All Electric Range (Miles) 72° HVAC OFF per EPA	18
All Electric Driving Range (Miles) Winter (AAA Study)	9
All Electric Driving Range (Miles) Summer (AAA Study)	10

Assumes Diesel and Electricity Prices do not change in relation to gasoline

BMW I8 Roasdster

MSRP: $ 163,300
Gasoline Counterpart: 840i Convertible

Lifetime Economic Costs	
Technology Cost	$ 65,900
Net Fuel Cost (Savings)	$ (543)
Net Taxes & Fees	$ 3,598
Net Financing	$ 10,096
Net Insurance	$ 12,673
Relative Value Loss	$ -
Net Cost (Savings)	$ 91,724

Social Benefits	
Refueling Time Savings	$ (5)
Energy Security	$ (7)
Social Cost of Carbon Emissions	$ 574
Crashes, Fatalities, Congestion, Noise	$ (5)
Lost Fuel Tax Revenue	$ (908)
Net Cost (Savings) Including Social Benefits	$ 92,074

Meets 2020 Target	✓
Meets Original 2025 Target	✗
Saves Money (including Social Benefits)	✗
Breakeven cost of Gasoline (per Gallon)	NA

All Electric Range (Miles) 72° HVAC OFF per EPA	18
All Electric Driving Range (Miles) Winter (AAA Study)	9
All Electric Driving Range (Miles) Summer (AAA Study)	10

Assumes Diesel and Electricity Prices do not change in relation to gasoline

Plug-in Hybrid Electric Vehicles

FCA Pacifica

MSRP: $ 40,245
Gasoline Counterpart: Pacifica

Lifetime Economic Costs		
Technology Cost	$	6,795
Net Fuel Cost (Savings)	$	(3,841)
Net Taxes & Fees	$	371
Net Financing	$	1,041
Net Insurance	$	1,307
Relative Value Loss	$	-
Net Cost (Savings)	$	5,672

Social Benefits		
Refueling Time Savings	$	20
Energy Security	$	28
Social Cost of Carbon Emissions	$	1,098
Crashes, Fatalities, Congestion, Noise	$	(32)
Lost Fuel Tax Revenue	$	(1,660)
Net Cost (Savings) Including Social Benefits	$	6,219

Meets 2020 Target	✓	
Meets Original 2025 Target	✓	
Saves Money (including Social Benefits)	✗	
Breakeven cost of Gasoline (per Gallon)	$	2.64

All Electric Range (Miles) 72° HVAC OFF per EPA	32
All Electric Driving Range (Miles) Winter (AAA Study)	15
All Electric Driving Range (Miles) Summer (AAA Study)	18

Assumes Diesel and Electricity Prices do not change in relation to gasoline

Plug-in Hybrid Electric Vehicles

Ford Fusion Energy

MSRP: $ 35,000
Gasoline Counterpart: Fusion

Lifetime Economic Costs	
Technology Cost	$ 11,830
Net Fuel Cost (Savings)	$ (3,293)
Net Taxes & Fees	$ 646
Net Financing	$ 1,812
Net Insurance	$ 2,275
Relative Value Loss	$ -
Net Cost (Savings)	$ 13,270

Social Benefits	
Refueling Time Savings	$ 40
Energy Security	$ 56
Social Cost of Carbon Emissions	$ 1,084
Crashes, Fatalities, Congestion, Noise	$ (28)
Lost Fuel Tax Revenue	$ (1,251)
Net Cost (Savings) Including Social Benefits	$ 13,369

Meets 2020 Target	✓
Meets Original 2025 Target	✓
Saves Money (including Social Benefits)	✗
Breakeven cost of Gasoline (per Gallon)	$ 5.57

All Electric Range (Miles) 72° HVAC OFF per EPA	26
All Electric Driving Range (Miles) Winter (AAA Study)	12
All Electric Driving Range (Miles) Summer (AAA Study)	14

Assumes Diesel and Electricity Prices do not change in relation to gasoline

Ford Fusion Special

MSRP: $ 35,000
Gasoline Counterpart: Fusion

Lifetime Economic Costs	
Technology Cost	$ 11,830
Net Fuel Cost (Savings)	$ (2,973)
Net Taxes & Fees	$ 646
Net Financing	$ 1,812
Net Insurance	$ 2,275
Relative Value Loss	$ -
Net Cost (Savings)	$ 13,590

Social Benefits	
Refueling Time Savings	$ 36
Energy Security	$ 51
Social Cost of Carbon Emissions	$ 1,026
Crashes, Fatalities, Congestion, Noise	$ (25)
Lost Fuel Tax Revenue	$ (1,200)
Net Cost (Savings) Including Social Benefits	$ 13,702

Meets 2020 Target	✓
Meets Original 2025 Target	✓
Saves Money (including Social Benefits)	✗
Breakeven cost of Gasoline (per Gallon)	$ 5.78

All Electric Range (Miles) 72° HVAC OFF per EPA	26
All Electric Driving Range (Miles) Winter (AAA Study)	12
All Electric Driving Range (Miles) Summer (AAA Study)	14

Assumes Diesel and Electricity Prices do not change in relation to gasoline

Hyundai Ioniq Plug-in Hybrid

MSRP: $ 25,305
Gasoline Counterpart: Ioniq

Lifetime Economic Costs		
Technology Cost	$	4,655
Net Fuel Cost (Savings)	$	(4,130)
Net Taxes & Fees	$	254
Net Financing	$	713
Net Insurance	$	895
Relative Value Loss	$	-
Net Cost (Savings)	$	2,387

Social Benefits		
Refueling Time Savings	$	(53)
Energy Security	$	(75)
Social Cost of Carbon Emissions	$	147
Crashes, Fatalities, Congestion, Noise	$	(35)
Lost Fuel Tax Revenue	$	(387)
Net Cost (Savings) Including Social Benefits	$	2,789

Meets 2020 Target	✓
Meets Original 2025 Target	✓
Saves Money (including Social Benefits)	✗
Breakeven cost of Gasoline (per Gallon)	NA

All Electric Range (Miles) 72° HVAC OFF per EPA	29
All Electric Driving Range (Miles) Winter (AAA Study)	14
All Electric Driving Range (Miles) Summer (AAA Study)	16

Assumes Diesel and Electricity Prices do not change in relation to gasoline

KIA Niro Plug-in Hybrid

MSRP: $ 28,500
Gasoline Counterpart: Niro

Lifetime Economic Costs		
Technology Cost	$	5,010
Net Fuel Cost (Savings)	$	(3,314)
Net Taxes & Fees	$	274
Net Financing	$	768
Net Insurance	$	963
Relative Value Loss	$	-
Net Cost (Savings)	$	3,701

Social Benefits		
Refueling Time Savings	$	(74)
Energy Security	$	(105)
Social Cost of Carbon Emissions	$	8
Crashes, Fatalities, Congestion, Noise	$	(28)
Lost Fuel Tax Revenue	$	(273)
Net Cost (Savings) Including Social Benefits	$	4,173

Meets 2020 Target	✓
Meets Original 2025 Target	✓
Saves Money (including Social Benefits)	✗
Breakeven cost of Gasoline (per Gallon)	NA

All Electric Range (Miles) 72° HVAC OFF per EPA	26
All Electric Driving Range (Miles) Winter (AAA Study)	12
All Electric Driving Range (Miles) Summer (AAA Study)	14

Assumes Diesel and Electricity Prices do not change in relation to gasoline

KIA Optima

MSRP: $ 36,090
Gasoline Counterpart: Optima

Lifetime Economic Costs		
Technology Cost	$	20,650
Net Fuel Cost (Savings)	$	(1,882)
Net Taxes & Fees	$	1,127
Net Financing	$	3,164
Net Insurance	$	3,971
Relative Value Loss	$	-
Net Cost (Savings)	$	27,030

Social Benefits		
Refueling Time Savings	$	18
Energy Security	$	26
Social Cost of Carbon Emissions	$	778
Crashes, Fatalities, Congestion, Noise	$	(16)
Lost Fuel Tax Revenue	$	(1,019)
Net Cost (Savings) Including Social Benefits	$	27,243

Meets 2020 Target	✓
Meets Original 2025 Target	✓
Saves Money (including Social Benefits)	✗
Breakeven cost of Gasoline (per Gallon)	$ 7.31

All Electric Range (Miles) 72° HVAC OFF per EPA	28
All Electric Driving Range (Miles) Winter (AAA Study)	13
All Electric Driving Range (Miles) Summer (AAA Study)	15

Lincoln Aviator Grand Tournig AWD

MSRP: $ 69,540
Gasoline Counterpart: Aviator Reserve AWD

Lifetime Economic Costs	
Technology Cost	$ 20,651
Net Fuel Cost (Savings)	$ (1,265)
Net Taxes & Fees	$ 1,128
Net Financing	$ 3,164
Net Insurance	$ 3,971
Relative Value Loss	$ -
Net Cost (Savings)	$ 27,648

Social Benefits	
Refueling Time Savings	$ 0
Energy Security	$ 0
Social Cost of Carbon Emissions	$ 733
Crashes, Fatalities, Congestion, Noise	$ (11)
Lost Fuel Tax Revenue	$ (1,368)
Net Cost (Savings) Including Social Benefits	$ 28,293

Meets 2020 Target	✓
Meets Original 2025 Target	✗
Saves Money (including Social Benefits)	✗
Breakeven cost of Gasoline (per Gallon)	$ 4.43

All Electric Range (Miles) 72° HVAC OFF per EPA	21
All Electric Driving Range (Miles) Winter (AAA Study)	10
All Electric Driving Range (Miles) Summer (AAA Study)	12

Assumes Diesel and Electricity Prices do not change in relation to gasoline

Mini MINI Cooper SE Countryman ALL4

MSRP: $ 36,900
Gasoline Counterpart: Countryman S All4

Lifetime Economic Costs	
Technology Cost	$ 20,652
Net Fuel Cost (Savings)	$ (80)
Net Taxes & Fees	$ 1,128
Net Financing	$ 3,164
Net Insurance	$ 3,971
Relative Value Loss	$ -
Net Cost (Savings)	$ 28,835

Social Benefits	
Refueling Time Savings	$ (9)
Energy Security	$ (12)
Social Cost of Carbon Emissions	$ 489
Crashes, Fatalities, Congestion, Noise	$ (1)
Lost Fuel Tax Revenue	$ (778)
Net Cost (Savings) Including Social Benefits	$ 29,145

Meets 2020 Target	✓
Meets Original 2025 Target	✗
Saves Money (including Social Benefits)	✗
Breakeven cost of Gasoline (per Gallon)	NA

All Electric Range (Miles) 72° HVAC OFF per EPA	18
All Electric Driving Range (Miles) Winter (AAA Study)	9
All Electric Driving Range (Miles) Summer (AAA Study)	10

Assumes Diesel and Electricity Prices do not change in relation to gasoline

Mitsubishi Outlander

MSRP: $ 36,095
Gasoline Counterpart: Outlander

Lifetime Economic Costs	
Technology Cost	$ 11,300
Net Fuel Cost (Savings)	$ (173)
Net Taxes & Fees	$ 617
Net Financing	$ 1,731
Net Insurance	$ 2,173
Relative Value Loss	$ -
Net Cost (Savings)	$ 15,648

Social Benefits	
Refueling Time Savings	$ (11)
Energy Security	$ (15)
Social Cost of Carbon Emissions	$ 443
Crashes, Fatalities, Congestion, Noise	$ (1)
Lost Fuel Tax Revenue	$ (895)
Net Cost (Savings) Including Social Benefits	$ 16,128

Meets 2020 Target	✓
Meets Original 2025 Target	✓
Saves Money (including Social Benefits)	✗
Breakeven cost of Gasoline (per Gallon)	NA

All Electric Range (Miles) 72° HVAC OFF per EPA	22
All Electric Driving Range (Miles) Winter (AAA Study)	10
All Electric Driving Range (Miles) Summer (AAA Study)	12

Assumes Diesel and Electricity Prices do not change in relation to gasoline

Plug-in Hybrid Electric Vehicles

Porsche Panamera 4 e-Hybrid

MSRP: $ 102,900
Gasoline Counterpart: Panamera 4

Lifetime Economic Costs	
Technology Cost	$ 16,600
Net Fuel Cost (Savings)	$ 1,158
Net Taxes & Fees	$ 906
Net Financing	$ 2,543
Net Insurance	$ 3,192
Relative Value Loss	$ -
Net Cost (Savings)	$ 24,400

Social Benefits	
Refueling Time Savings	$ (11)
Energy Security	$ (16)
Social Cost of Carbon Emissions	$ 373
Crashes, Fatalities, Congestion, Noise	$ 10
Lost Fuel Tax Revenue	$ (789)
Net Cost (Savings) Including Social Benefits	$ 24,833

Meets 2020 Target	✓
Meets Original 2025 Target	✓
Saves Money (including Social Benefits)	✗
Breakeven cost of Gasoline (per Gallon)	NA

All Electric Range (Miles) 72° HVAC OFF per EPA	14
All Electric Driving Range (Miles) Winter (AAA Study)	7
All Electric Driving Range (Miles) Summer (AAA Study)	8

Assumes Diesel and Electricity Prices do not change in relation to gasoline

Porsche Panamera 4 e-Hybrid Executive

MSRP: $ 108,300
Gasoline Counterpart: Panamera 4 Executive

Lifetime Economic Costs		
Technology Cost	$	9,800
Net Fuel Cost (Savings)	$	1,158
Net Taxes & Fees	$	535
Net Financing	$	1,501
Net Insurance	$	1,885
Relative Value Loss	$	-
Net Cost (Savings)	$	14,879

Social Benefits		
Refueling Time Savings	$	(11)
Energy Security	$	(16)
Social Cost of Carbon Emissions	$	373
Crashes, Fatalities, Congestion, Noise	$	10
Lost Fuel Tax Revenue	$	(789)
Net Cost (Savings) Including Social Benefits		$ 15,312

Meets 2020 Target	✓
Meets Original 2025 Target	✗
Saves Money (including Social Benefits)	✗
Breakeven cost of Gasoline (per Gallon)	NA

All Electric Range (Miles) 72° HVAC OFF per EPA	14
All Electric Driving Range (Miles) Winter (AAA Study)	7
All Electric Driving Range (Miles) Summer (AAA Study)	8

Assumes Diesel and Electricity Prices do not change in relation to gasoline

Porsche Panamera 4 e-Hybrid ST

MSRP: $ 107,800
Gasoline Counterpart: Panamera 4 ST

Lifetime Economic Costs		
Technology Cost	$	9,800
Net Fuel Cost (Savings)	$	940
Net Taxes & Fees	$	535
Net Financing	$	1,501
Net Insurance	$	1,885
Relative Value Loss	$	-
Net Cost (Savings)	$	14,661

Social Benefits		
Refueling Time Savings	$	(10)
Energy Security	$	(14)
Social Cost of Carbon Emissions	$	419
Crashes, Fatalities, Congestion, Noise	$	8
Lost Fuel Tax Revenue	$	(830)
Net Cost (Savings) Including Social Benefits	$	15,087

Meets 2020 Target	✓
Meets Original 2025 Target	✗
Saves Money (including Social Benefits)	✗
Breakeven cost of Gasoline (per Gallon)	NA

All Electric Range (Miles) 72° HVAC OFF per EPA	14
All Electric Driving Range (Miles) Winter (AAA Study)	7
All Electric Driving Range (Miles) Summer (AAA Study)	8

Assumes Diesel and Electricity Prices do not change in relation to gasoline

Plug-in Hybrid Electric Vehicles

Porsche Panamera Turbo S e-Hybrid

MSRP: $ 187,700
Gasoline Counterpart: Panamera Turbo S

Lifetime Economic Costs	
Technology Cost	$ 34,700
Net Fuel Cost (Savings)	$ 1,254
Net Taxes & Fees	$ 1,895
Net Financing	$ 5,316
Net Insurance	$ 6,673
Relative Value Loss	$ -
Net Cost (Savings)	$ 49,838

Social Benefits	
Refueling Time Savings	$ (13)
Energy Security	$ (18)
Social Cost of Carbon Emissions	$ 275
Crashes, Fatalities, Congestion, Noise	$ 11
Lost Fuel Tax Revenue	$ (791)
Net Cost (Savings) Including Social Benefits	$ 50,374

Meets 2020 Target	✓
Meets Original 2025 Target	✗
Saves Money (including Social Benefits)	✗
Breakeven cost of Gasoline (per Gallon)	NA

All Electric Range (Miles) 72° HVAC OFF per EPA	14
All Electric Driving Range (Miles) Winter (AAA Study)	7
All Electric Driving Range (Miles) Summer (AAA Study)	8

Assumes Diesel and Electricity Prices do not change in relation to gasoline

Porsche Panamera Turbo S e-Hybrid Exec

MSRP: $ 198,100
Gasoline Counterpart: Panamera Turbo Executive

Lifetime Economic Costs	
Technology Cost	$ 34,700
Net Fuel Cost (Savings)	$ 1,362
Net Taxes & Fees	$ 1,895
Net Financing	$ 5,316
Net Insurance	$ 6,673
Relative Value Loss	$ -
Net Cost (Savings)	$ 49,945

Social Benefits	
Refueling Time Savings	$ (13)
Energy Security	$ (18)
Social Cost of Carbon Emissions	$ 275
Crashes, Fatalities, Congestion, Noise	$ 11
Lost Fuel Tax Revenue	$ (791)
Net Cost (Savings) Including Social Benefits	$ 50,481

Meets 2020 Target	✓
Meets Original 2025 Target	✗
Saves Money (including Social Benefits)	✗
Breakeven cost of Gasoline (per Gallon)	NA

All Electric Range (Miles) 72° HVAC OFF per EPA	14
All Electric Driving Range (Miles) Winter (AAA Study)	7
All Electric Driving Range (Miles) Summer (AAA Study)	8

Assumes Diesel and Electricity Prices do not change in relation to gasoline

Porsche Panamera Turbo S e-Hybrid ST

MSRP: $ 191,700
Gasoline Counterpart: Panamera Turbo ST

Lifetime Economic Costs		
Technology Cost	$	34,700
Net Fuel Cost (Savings)	$	649
Net Taxes & Fees	$	1,895
Net Financing	$	5,316
Net Insurance	$	6,673
Relative Value Loss	$	-
Net Cost (Savings)	$	49,233

Social Benefits		
Refueling Time Savings	$	(8)
Energy Security	$	(11)
Social Cost of Carbon Emissions	$	446
Crashes, Fatalities, Congestion, Noise	$	5
Lost Fuel Tax Revenue	$	(944)
Net Cost (Savings) Including Social Benefits	$	49,744

Meets 2020 Target	✓
Meets Original 2025 Target	✗
Saves Money (including Social Benefits)	✗
Breakeven cost of Gasoline (per Gallon)	NA

All Electric Range (Miles) 72° HVAC OFF per EPA	14
All Electric Driving Range (Miles) Winter (AAA Study)	7
All Electric Driving Range (Miles) Summer (AAA Study)	8

Assumes Diesel and Electricity Prices do not change in relation to gasoline

Plug-in Hybrid Electric Vehicles

Subaru Crosstrek AWD

MSRP: $ 34,995
Gasoline Counterpart: Crosstrek AWD

Lifetime Economic Costs		
Technology Cost	$	13,100
Net Fuel Cost (Savings)	$	(1,182)
Net Taxes & Fees	$	715
Net Financing	$	2,007
Net Insurance	$	2,519
Relative Value Loss	$	-
Net Cost (Savings)	$	17,159

Social Benefits		
Refueling Time Savings	$	6
Energy Security	$	8
Social Cost of Carbon Emissions	$	667
Crashes, Fatalities, Congestion, Noise	$	(10)
Lost Fuel Tax Revenue	$	(895)
Net Cost (Savings) Including Social Benefits	$	17,383

Meets 2020 Target	✓
Meets Original 2025 Target	✓
Saves Money (including Social Benefits)	✗
Breakeven cost of Gasoline (per Gallon)	$ 8.39

All Electric Range (Miles) 72° HVAC OFF per EPA	17
All Electric Driving Range (Miles) Winter (AAA Study)	8
All Electric Driving Range (Miles) Summer (AAA Study)	9

Assumes Diesel and Electricity Prices do not change in relation to gasoline

Plug-in Hybrid Electric Vehicles

Toyota Prius Prime

MSRP: $ 27,600
Gasoline Counterpart: Yaris

Lifetime Economic Costs	
Technology Cost	$ 11,950
Net Fuel Cost (Savings)	$ (2,124)
Net Taxes & Fees	$ 652
Net Financing	$ 1,831
Net Insurance	$ 2,298
Relative Value Loss	$ -
Net Cost (Savings)	$ 14,607

Social Benefits	
Refueling Time Savings	$ 39
Energy Security	$ 55
Social Cost of Carbon Emissions	$ 783
Crashes, Fatalities, Congestion, Noise	$ (18)
Lost Fuel Tax Revenue	$ (897)
Net Cost (Savings) Including Social Benefits	$ 14,645

Meets 2020 Target	✓
Meets Original 2025 Target	✓
Saves Money (including Social Benefits)	✗
Breakeven cost of Gasoline (per Gallon)	$ 7.67

All Electric Range (Miles) 72° HVAC OFF per EPA	25
All Electric Driving Range (Miles) Winter (AAA Study)	12
All Electric Driving Range (Miles) Summer (AAA Study)	14

Assumes Diesel and Electricity Prices do not change in relation to gasoline

Volvo S60

MSRP: $ 55,400
Gasoline Counterpart: S60

Lifetime Economic Costs	
Technology Cost	$ 19,350
Net Fuel Cost (Savings)	$ (309)
Net Taxes & Fees	$ 1,057
Net Financing	$ 2,964
Net Insurance	$ 3,721
Relative Value Loss	$ -
Net Cost (Savings)	$ 26,783

Social Benefits	
Refueling Time Savings	$ (6)
Energy Security	$ (8)
Social Cost of Carbon Emissions	$ 537
Crashes, Fatalities, Congestion, Noise	$ (3)
Lost Fuel Tax Revenue	$ (875)
Net Cost (Savings) Including Social Benefits	$ 27,137

Meets 2020 Target	✓
Meets Original 2025 Target	✓
Saves Money (including Social Benefits)	✗
Breakeven cost of Gasoline (per Gallon)	NA

All Electric Range (Miles) 72° HVAC OFF per EPA	22
All Electric Driving Range (Miles) Winter (AAA Study)	10
All Electric Driving Range (Miles) Summer (AAA Study)	12

Assumes Diesel and Electricity Prices do not change in relation to gasoline

Volvo S90

MSRP: $ 63,900
Gasoline Counterpart: S90

Lifetime Economic Costs	
Technology Cost	$ 16,550
Net Fuel Cost (Savings)	$ (961)
Net Taxes & Fees	$ 904
Net Financing	$ 2,535
Net Insurance	$ 3,183
Relative Value Loss	$ -
Net Cost (Savings)	$ 22,210

Social Benefits	
Refueling Time Savings	$ 4
Energy Security	$ 5
Social Cost of Carbon Emissions	$ 739
Crashes, Fatalities, Congestion, Noise	$ (8)
Lost Fuel Tax Revenue	$ (1,049)

Net Cost (Savings) Including Social Benefits	$ 22,520

Meets 2020 Target	✓
Meets Original 2025 Target	✓
Saves Money (including Social Benefits)	✗
Breakeven cost of Gasoline (per Gallon)	$ 9.01

All Electric Range (Miles) 72° HVAC OFF per EPA	21
All Electric Driving Range (Miles) Winter (AAA Study)	10
All Electric Driving Range (Miles) Summer (AAA Study)	12

Assumes Diesel and Electricity Prices do not change in relation to gasoline

Plug-in Hybrid Electric Vehicles

Volvo V60

MSRP: $ 67,300
Gasoline Counterpart: V60

Lifetime Economic Costs	
Technology Cost	$ 21,555
Net Fuel Cost (Savings)	$ (337)
Net Taxes & Fees	$ 1,177
Net Financing	$ 3,302
Net Insurance	$ 4,145
Relative Value Loss	$ -
Net Cost (Savings)	$ 29,842

Social Benefits	
Refueling Time Savings	$ (6)
Energy Security	$ (8)
Social Cost of Carbon Emissions	$ 537
Crashes, Fatalities, Congestion, Noise	$ (3)
Lost Fuel Tax Revenue	$ (955)
Net Cost (Savings) Including Social Benefits	$ 30,277

Meets 2020 Target	✓
Meets Original 2025 Target	✓
Saves Money (including Social Benefits)	✗
Breakeven cost of Gasoline (per Gallon)	NA

All Electric Range (Miles) 72° HVAC OFF per EPA	22
All Electric Driving Range (Miles) Winter (AAA Study)	10
All Electric Driving Range (Miles) Summer (AAA Study)	12

Assumes Diesel and Electricity Prices do not change in relation to gasoline

Volvo XC60

MSRP: $ 54,595
Gasoline Counterpart: XC60

Lifetime Economic Costs	
Technology Cost	$ 13,800
Net Fuel Cost (Savings)	$ (606)
Net Taxes & Fees	$ 753
Net Financing	$ 2,114
Net Insurance	$ 2,654
Relative Value Loss	$ -
Net Cost (Savings)	$ 18,715

Social Benefits	
Refueling Time Savings	$ 0
Energy Security	$ 0
Social Cost of Carbon Emissions	$ 693
Crashes, Fatalities, Congestion, Noise	$ (5)
Lost Fuel Tax Revenue	$ (1,141)
Net Cost (Savings) Including Social Benefits	$ 19,168

Meets 2020 Target	✓
Meets Original 2025 Target	✓
Saves Money (including Social Benefits)	✗
Breakeven cost of Gasoline (per Gallon)	$ 7.00

All Electric Range (Miles) 72° HVAC OFF per EPA	19
All Electric Driving Range (Miles) Winter (AAA Study)	9
All Electric Driving Range (Miles) Summer (AAA Study)	10

Assumes Diesel and Electricity Prices do not change in relation to gasoline

Plug-in Hybrid Electric Vehicles

Volvo XC90

MSRP: $ 67,500
Gasoline Counterpart: XC90

Lifetime Economic Costs	
Technology Cost	$ 19,150
Net Fuel Cost (Savings)	$ 264
Net Taxes & Fees	$ 1,046
Net Financing	$ 2,934
Net Insurance	$ 3,683
Relative Value Loss	$ -
Net Cost (Savings)	$ 27,076

Social Benefits	
Refueling Time Savings	$ (6)
Energy Security	$ (8)
Social Cost of Carbon Emissions	$ 553
Crashes, Fatalities, Congestion, Noise	$ 2
Lost Fuel Tax Revenue	$ (978)
Net Cost (Savings) Including Social Benefits	$ 27,513

Meets 2020 Target	✓
Meets Original 2025 Target	✓
Saves Money (including Social Benefits)	✗
Breakeven cost of Gasoline (per Gallon)	NA

All Electric Range (Miles) 72° HVAC OFF per EPA	18
All Electric Driving Range (Miles) Winter (AAA Study)	9
All Electric Driving Range (Miles) Summer (AAA Study)	10

Assumes Diesel and Electricity Prices do not change in relation to gasoline

Karma Revero GT

MSRP: $ 135,000
Gasoline Counterpart: Average Non PHEV

Lifetime Economic Costs	
Technology Cost	$ 16,000
Net Fuel Cost (Savings)	$ (196)
Net Taxes & Fees	$ 874
Net Financing	$ 2,451
Net Insurance	$ 3,077
Relative Value Loss	$ -
Net Cost (Savings)	$ 22,206

Social Benefits	
Refueling Time Savings	$ (22)
Energy Security	$ (32)
Social Cost of Carbon Emissions	$ 192
Crashes, Fatalities, Congestion, Noise	$ (2)
Lost Fuel Tax Revenue	$ (1,579)
Net Cost (Savings) Including Social Benefits	$ 23,649

Meets 2020 Target	✓
Meets Original 2025 Target	✓
Saves Money (including Social Benefits)	✗
Breakeven cost of Gasoline (per Gallon)	NA

All Electric Range (Miles) 72° HVAC OFF per EPA	61
All Electric Driving Range (Miles) Winter (AAA Study)	29
All Electric Driving Range (Miles) Summer (AAA Study)	34

Assumes Diesel and Electricity Prices do not change in relation to gasoline

Plug-in Hybrid Electric Vehicles

Land Rover Range Rover Sport PHEV

MSRP: $ 79,000
Gasoline Counterpart: Range Rover Sport

Lifetime Economic Costs	
Technology Cost	$ 10,350
Net Fuel Cost (Savings)	$ 1,282
Net Taxes & Fees	$ 565
Net Financing	$ 1,586
Net Insurance	$ 1,990
Relative Value Loss	$ -
Net Cost (Savings)	$ 15,773

Social Benefits	
Refueling Time Savings	$ (7)
Energy Security	$ (11)
Social Cost of Carbon Emissions	$ 418
Crashes, Fatalities, Congestion, Noise	$ 11
Lost Fuel Tax Revenue	$ (1,212)
Net Cost (Savings) Including Social Benefits	$ 16,573

Meets 2020 Target	✓
Meets Original 2025 Target	✓
Saves Money (including Social Benefits)	✗
Breakeven cost of Gasoline (per Gallon)	NA

All Electric Range (Miles) 72° HVAC OFF per EPA	19
All Electric Driving Range (Miles) Winter (AAA Study)	9
All Electric Driving Range (Miles) Summer (AAA Study)	10

Assumes Diesel and Electricity Prices do not change in relation to gasoline

Land Rover Range Rover PHEV

MSRP: $ 95,950
Gasoline Counterpart: Range Rover

Lifetime Economic Costs	
Technology Cost	$ 5,050
Net Fuel Cost (Savings)	$ 774
Net Taxes & Fees	$ 276
Net Financing	$ 774
Net Insurance	$ 971
Relative Value Loss	$ -
Net Cost (Savings)	$ 7,844

Social Benefits	
Refueling Time Savings	$ (5)
Energy Security	$ (7)
Social Cost of Carbon Emissions	$ 515
Crashes, Fatalities, Congestion, Noise	$ 7
Lost Fuel Tax Revenue	$ (1,306)
Net Cost (Savings) Including Social Benefits	$ 8,641

Meets 2020 Target	✓
Meets Original 2025 Target	✓
Saves Money (including Social Benefits)	✗
Breakeven cost of Gasoline (per Gallon)	NA

All Electric Range (Miles) 72° HVAC OFF per EPA	19
All Electric Driving Range (Miles) Winter (AAA Study)	9
All Electric Driving Range (Miles) Summer (AAA Study)	10

Assumes Diesel and Electricity Prices do not change in relation to gasoline

HYBRID ELECTRIC VEHICLES

The 2020 Model Year EPA/NHTSA Fuel Economy Guide contains information on seventy-two hybrid electric vehicles. **Fifty-one percent of these vehicles meet their 2020 model year fuel economy targets** and thirty-nine percent meet their 2025 model year fuel economy targets.

Eleven percent (8 of 72) saved the customer money using the methodology employed by the National Highway Traffic Safety Administration and EPA for calculating the cost and benefits of technology.

The average cost premium for a hybrid electric vehicle was $6500.

Readers will note that the guide reports the on-road fuel economy of hybrid electric vehicles and diesel vehicles. The information is based on data reported by the US Environmental Protection Agency in their annual test car dataset. This dataset reflects testing conducted using the so-called five-cycle criteria. Each year the agency tests a random sample of vehicles using five unique test cycles that reflect different driving styles and environmental conditions including ambient temperature and cabin heating and cooling. CAFE testing is conducted at 72°F with all accessories including heat and air conditioning turned off. As shown in the tables hot summer days with the air conditioning turned on dramatically affects fuel economy according to EPA. The same is true for cold winter days when the heater is turned on. The 'typical driving' cycle is one that EPA developed to more accurately represent vehicle acceleration rates and city driving speeds.

Jeep Wrangler Unlimited 4x4

MSRP: $ 33,295
Gasoline Counterpart: Wrangler Unlimited 4x4

Lifetime Economic Costs		
Technology Cost	$	1,500
Net Fuel Cost (Savings)	$	2,564
Net Taxes & Fees	$	82
Net Financing	$	230
Net Insurance	$	288
Relative Value Loss	$	-
Net Cost (Savings)	$	4,664

Social Benefits		
Refueling Time Savings	$	(17)
Energy Security	$	(24)
Social Cost of Carbon Emissions	$	-
Crashes, Fatalities, Congestion, Noise	$	-
Lost Fuel Tax Revenue	$	479
Net Cost (Savings) Including Social Benefits	$	4,226

Meets 2020 Target	✓
Meets Original 2025 Target	✓
Saves Money (including Social Benefits)	✗
Breakeven cost of Gasoline (per Gallon)	$ -

CAFE Fuel Economy (mpg)	30
On Road mpg Winter (EPA 5-cycle Study)	22
On Road mpg Summer (EPA 5-cycle Study)	24
On Road mpg Typical Driving - Accessories OFF (EPA 5-cycle Study)	26

Jeep Wrangler Unlimited 4x4

MSRP: $ 34,295
Gasoline Counterpart: Wrangler Unlimited 4x4

Lifetime Economic Costs		
Technology Cost	$	2,500
Net Fuel Cost (Savings)	$	3,833
Net Taxes & Fees	$	137
Net Financing	$	383
Net Insurance	$	481
Relative Value Loss	$	-
Net Cost (Savings)	$	7,333

Social Benefits		
Refueling Time Savings	$	(25)
Energy Security	$	(35)
Social Cost of Carbon Emissions	$	-
Crashes, Fatalities, Congestion, Noise	$	-
Lost Fuel Tax Revenue	$	716
Net Cost (Savings) Including Social Benefits	$	6,678

Meets 2020 Target	✓
Meets Original 2025 Target	✓
Saves Money (including Social Benefits)	✗
Breakeven cost of Gasoline (per Gallon)	$ -

CAFE Fuel Economy (mpg)	27
On Road mpg Winter (EPA 5-cycle Study)	20
On Road mpg Summer (EPA 5-cycle Study)	22
On Road mpg Typical Driving - Accessories OFF (EPA 5-cycle Study)	24

RAM 1500 4X2

MSRP: $ 38,785
Gasoline Counterpart: 1500 4X2

Lifetime Economic Costs		
Technology Cost	$	200
Net Fuel Cost (Savings)	$	(1,393)
Net Taxes & Fees	$	11
Net Financing	$	31
Net Insurance	$	38
Relative Value Loss	$	-
Net Cost (Savings)	$	(1,113)

Social Benefits		
Refueling Time Savings	$	7
Energy Security	$	10
Social Cost of Carbon Emissions	$	-
Crashes, Fatalities, Congestion, Noise	$	-
Lost Fuel Tax Revenue	$	(260)
Net Cost (Savings) Including Social Benefits	$	(870)

Meets 2020 Target	✓
Meets Original 2025 Target	✓
Saves Money (including Social Benefits)	✓
Breakeven cost of Gasoline (per Gallon)	$ -

CAFE Fuel Economy (mpg)	29
On Road mpg Winter (EPA 5-cycle Study)	22
On Road mpg Summer (EPA 5-cycle Study)	23
On Road mpg Typical Driving - Accessories OFF (EPA 5-cycle Study)	25

RAM 1500 4X2

MSRP: $ 38,785
Gasoline Counterpart: 1500 4X2

Lifetime Economic Costs		
Technology Cost	$	200
Net Fuel Cost (Savings)	$	507
Net Taxes & Fees	$	11
Net Financing	$	31
Net Insurance	$	38
Relative Value Loss	$	-
Net Cost (Savings)	$	787

Social Benefits		
Refueling Time Savings	$	(2)
Energy Security	$	(3)
Social Cost of Carbon Emissions	$	-
Crashes, Fatalities, Congestion, Noise	$	-
Lost Fuel Tax Revenue	$	95
Net Cost (Savings) Including Social Benefits	$	698

Meets 2020 Target	✓
Meets Original 2025 Target	✓
Saves Money (including Social Benefits)	✗
Breakeven cost of Gasoline (per Gallon)	$ -

CAFE Fuel Economy (mpg)	25
On Road mpg Winter (EPA 5-cycle Study)	19
On Road mpg Summer (EPA 5-cycle Study)	21
On Road mpg Typical Driving - Accessories OFF (EPA 5-cycle Study)	23

RAM 1500 4X4

MSRP: $ 49,306
Gasoline Counterpart: 1500 4X4

Lifetime Economic Costs	
Technology Cost	$ 1,250
Net Fuel Cost (Savings)	$ (1,246)
Net Taxes & Fees	$ 68
Net Financing	$ 192
Net Insurance	$ 240
Relative Value Loss	$ -
Net Cost (Savings)	$ 504

Social Benefits	
Refueling Time Savings	$ 6
Energy Security	$ 9
Social Cost of Carbon Emissions	$ -
Crashes, Fatalities, Congestion, Noise	$ -
Lost Fuel Tax Revenue	$ (233)
Net Cost (Savings) Including Social Benefits	$ 722

Meets 2020 Target	✓
Meets Original 2025 Target	✓
Saves Money (including Social Benefits)	✗
Breakeven cost of Gasoline (per Gallon)	$ -

CAFE Fuel Economy (mpg)	28
On Road mpg Winter (EPA 5-cycle Study)	21
On Road mpg Summer (EPA 5-cycle Study)	23
On Road mpg Typical Driving - Accessories OFF (EPA 5-cycle Study)	24

RAM 1500 4X4

MSRP: $ 49,306
Gasoline Counterpart: 1500 4X4

Lifetime Economic Costs		
Technology Cost	$	1,250
Net Fuel Cost (Savings)	$	(1,419)
Net Taxes & Fees	$	68
Net Financing	$	192
Net Insurance	$	240
Relative Value Loss	$	-
Net Cost (Savings)	$	331

Social Benefits		
Refueling Time Savings	$	7
Energy Security	$	10
Social Cost of Carbon Emissions	$	-
Crashes, Fatalities, Congestion, Noise	$	-
Lost Fuel Tax Revenue	$	(265)
Net Cost (Savings) Including Social Benefits	$	579

Meets 2020 Target	✓
Meets Original 2025 Target	✓
Saves Money (including Social Benefits)	✗
Breakeven cost of Gasoline (per Gallon)	$ -

CAFE Fuel Economy (mpg)	24
On Road mpg Winter (EPA 5-cycle Study)	19
On Road mpg Summer (EPA 5-cycle Study)	20
On Road mpg Typical Driving - Accessories OFF (EPA 5-cycle Study)	23

RAM 1500 HFE 4X2

MSRP: $ 38,785
Gasoline Counterpart: 1500 HFE 4X2

Lifetime Economic Costs	
Technology Cost	$ 200
Net Fuel Cost (Savings)	$ (1,856)
Net Taxes & Fees	$ 11
Net Financing	$ 31
Net Insurance	$ 38
Relative Value Loss	$ -
Net Cost (Savings)	$ (1,576)

Social Benefits	
Refueling Time Savings	$ 10
Energy Security	$ 14
Social Cost of Carbon Emissions	$ -
Crashes, Fatalities, Congestion, Noise	$ -
Lost Fuel Tax Revenue	$ (346)
Net Cost (Savings) Including Social Benefits	$ (1,253)

Meets 2020 Target	✓
Meets Original 2025 Target	✓
Saves Money (including Social Benefits)	✓
Breakeven cost of Gasoline (per Gallon)	$ -

CAFE Fuel Economy (mpg)	30
On Road mpg Winter (EPA 5-cycle Study)	22
On Road mpg Summer (EPA 5-cycle Study)	24
On Road mpg Typical Driving - Accessories OFF (EPA 5-cycle Study)	25

Ford ESCAPE AWD HEV

MSRP: $ 31,095
Gasoline Counterpart: ESCAPE AWD

Lifetime Economic Costs		
Technology Cost	$	3,370
Net Fuel Cost (Savings)	$	(1,675)
Net Taxes & Fees	$	184
Net Financing	$	516
Net Insurance	$	648
Relative Value Loss	$	-
Net Cost (Savings)	$	3,043

Social Benefits		
Refueling Time Savings	$	23
Energy Security	$	33
Social Cost of Carbon Emissions	$	865
Crashes, Fatalities, Congestion, Noise	$	(14)
Lost Fuel Tax Revenue	$	(313)
Net Cost (Savings) Including Social Benefits	$	2,449

Meets 2020 Target	✓	
Meets Original 2025 Target	✓	
Saves Money (including Social Benefits)	✗	
Breakeven cost of Gasoline (per Gallon)	$	6.28

CAFE Fuel Economy (mpg)		57
On Road mpg Winter (EPA 5-cycle Study)		40
On Road mpg Summer (EPA 5-cycle Study)		41
On Road mpg Typical Driving - Accessories OFF (EPA 5-cycle Study)		40

Ford ESCAPE FWD HEV

MSRP: $ 29,595
Gasoline Counterpart: ESCAPE FWD

Lifetime Economic Costs	
Technology Cost	$ 3,370
Net Fuel Cost (Savings)	$ (1,441)
Net Taxes & Fees	$ 184
Net Financing	$ 516
Net Insurance	$ 648
Relative Value Loss	$ -
Net Cost (Savings)	$ 3,278

Social Benefits	
Refueling Time Savings	$ 21
Energy Security	$ 30
Social Cost of Carbon Emissions	$ 812
Crashes, Fatalities, Congestion, Noise	$ (12)
Lost Fuel Tax Revenue	$ (269)
Net Cost (Savings) Including Social Benefits	$ 2,696

Meets 2020 Target	✓
Meets Original 2025 Target	✓
Saves Money (including Social Benefits)	✗
Breakeven cost of Gasoline (per Gallon)	$ 7.24

CAFE Fuel Economy (mpg)	58
On Road mpg Winter (EPA 5-cycle Study)	41
On Road mpg Summer (EPA 5-cycle Study)	41
On Road mpg Typical Driving - Accessories OFF (EPA 5-cycle Study)	40

Ford EXPLORER HEV AWD

MSRP: $ 2,195
Gasoline Counterpart: EXPLORER LIMITED AWD

Lifetime Economic Costs		
Technology Cost	$	-
Net Fuel Cost (Savings)	$	1,083
Net Taxes & Fees	$	-
Net Financing	$	-
Net Insurance	$	-
Relative Value Loss	$	-
Net Cost (Savings)	$	1,083

Social Benefits		
Refueling Time Savings	$	(8)
Energy Security	$	(11)
Social Cost of Carbon Emissions	$	-
Crashes, Fatalities, Congestion, Noise	$	-
Lost Fuel Tax Revenue	$	202
Net Cost (Savings) Including Social Benefits	$	900

Meets 2020 Target	✓
Meets Original 2025 Target	✓
Saves Money (including Social Benefits)	✗
Breakeven cost of Gasoline (per Gallon)	$ -

CAFE Fuel Economy (mpg)	33
On Road mpg Winter (EPA 5-cycle Study)	24
On Road mpg Summer (EPA 5-cycle Study)	26
On Road mpg Typical Driving - Accessories OFF (EPA 5-cycle Study)	27

Ford EXPLORER HEV RWD

MSRP: $ 52,370
Gasoline Counterpart: EXPLORER LIMITED RWD

Lifetime Economic Costs		
Technology Cost	$	4,000
Net Fuel Cost (Savings)	$	169
Net Taxes & Fees	$	218
Net Financing	$	613
Net Insurance	$	769
Relative Value Loss	$	-
Net Cost (Savings)	$	5,770

Social Benefits		
Refueling Time Savings	$	(1)
Energy Security	$	(2)
Social Cost of Carbon Emissions	$	-
Crashes, Fatalities, Congestion, Noise	$	-
Lost Fuel Tax Revenue	$	32
Net Cost (Savings) Including Social Benefits	$	5,742

Meets 2020 Target	✓
Meets Original 2025 Target	✓
Saves Money (including Social Benefits)	✗
Breakeven cost of Gasoline (per Gallon)	$ -

CAFE Fuel Economy (mpg)	37
On Road mpg Winter (EPA 5-cycle Study)	27
On Road mpg Summer (EPA 5-cycle Study)	29
On Road mpg Typical Driving - Accessories OFF (EPA 5-cycle Study)	29

Ford FUSION HYBRID FWD

MSRP: $ 28,000
Gasoline Counterpart: FUSION HYBRID FWD

Lifetime Economic Costs	
Technology Cost	$ 4,830
Net Fuel Cost (Savings)	$ (2,887)
Net Taxes & Fees	$ 264
Net Financing	$ 740
Net Insurance	$ 929
Relative Value Loss	$ -
Net Cost (Savings)	$ 3,875

Social Benefits	
Refueling Time Savings	$ 44
Energy Security	$ 62
Social Cost of Carbon Emissions	$ 1,119
Crashes, Fatalities, Congestion, Noise	$ (24)
Lost Fuel Tax Revenue	$ (539)
Net Cost (Savings) Including Social Benefits	$ 3,214

Meets 2020 Target	✓
Meets Original 2025 Target	✓
Saves Money (including Social Benefits)	✗
Breakeven cost of Gasoline (per Gallon)	$ 5.29

CAFE Fuel Economy (mpg)	62
On Road mpg Winter (EPA 5-cycle Study)	44
On Road mpg Summer (EPA 5-cycle Study)	44
On Road mpg Typical Driving - Accessories OFF (EPA 5-cycle Study)	42

Ford FUSION HYBRID TAXI

MSRP: $ 28,000
Gasoline Counterpart: FUSION HYBRID TAXI

Lifetime Economic Costs		
Technology Cost	$	4,830
Net Fuel Cost (Savings)	$	(2,887)
Net Taxes & Fees	$	264
Net Financing	$	740
Net Insurance	$	929
Relative Value Loss	$	-
Net Cost (Savings)	$	3,875

Social Benefits		
Refueling Time Savings	$	44
Energy Security	$	62
Social Cost of Carbon Emissions	$	1,119
Crashes, Fatalities, Congestion, Noise	$	(24)
Lost Fuel Tax Revenue	$	(539)
Net Cost (Savings) Including Social Benefits	$	3,214

Meets 2020 Target	✓
Meets Original 2025 Target	✓
Saves Money (including Social Benefits)	✗
Breakeven cost of Gasoline (per Gallon)	$ 5.29

CAFE Fuel Economy (mpg)	62
On Road mpg Winter (EPA 5-cycle Study)	44
On Road mpg Summer (EPA 5-cycle Study)	44
On Road mpg Typical Driving - Accessories OFF (EPA 5-cycle Study)	42

Lincoln MKZ HYBRID FWD

MSRP: $ 35,955
Gasoline Counterpart: MKZ HYBRID FWD

Lifetime Economic Costs	
Technology Cost	$ -
Net Fuel Cost (Savings)	$ (4,385)
Net Taxes & Fees	$ -
Net Financing	$ -
Net Insurance	$ -
Relative Value Loss	$ -
Net Cost (Savings)	$ (4,385)

Social Benefits	
Refueling Time Savings	$ 58
Energy Security	$ 83
Social Cost of Carbon Emissions	$ 1,432
Crashes, Fatalities, Congestion, Noise	$ (37)
Lost Fuel Tax Revenue	$ (819)
Net Cost (Savings) Including Social Benefits	$ (5,102)

Meets 2020 Target	✓
Meets Original 2025 Target	✓
Saves Money (including Social Benefits)	✓
Breakeven cost of Gasoline (per Gallon)	$ 0.41

CAFE Fuel Economy (mpg)	62
On Road mpg Winter (EPA 5-cycle Study)	44
On Road mpg Summer (EPA 5-cycle Study)	44
On Road mpg Typical Driving - Accessories OFF (EPA 5-cycle Study)	42

Acura MDX AWD

MSRP: $ 52,900
Gasoline Counterpart: MDX AWD

Lifetime Economic Costs		
Technology Cost	$	3,500
Net Fuel Cost (Savings)	$	(715)
Net Taxes & Fees	$	191
Net Financing	$	536
Net Insurance	$	673
Relative Value Loss	$	-
Net Cost (Savings)	$	4,185

Social Benefits		
Refueling Time Savings	$	5
Energy Security	$	7
Social Cost of Carbon Emissions	$	823
Crashes, Fatalities, Congestion, Noise	$	(6)
Lost Fuel Tax Revenue	$	(133)
Net Cost (Savings) Including Social Benefits	$	3,489

Meets 2020 Target	✓
Meets Original 2025 Target	✗
Saves Money (including Social Benefits)	✗
Breakeven cost of Gasoline (per Gallon)	$ 14.71

CAFE Fuel Economy (mpg)	36
On Road mpg Winter (EPA 5-cycle Study)	26
On Road mpg Summer (EPA 5-cycle Study)	28
On Road mpg Typical Driving - Accessories OFF (EPA 5-cycle Study)	29

Acura NSX

MSRP: $ 159,495
Gasoline Counterpart: Average non-HEV Differential

Lifetime Economic Costs		
Technology Cost	$	7,100
Net Fuel Cost (Savings)	$	(76)
Net Taxes & Fees	$	388
Net Financing	$	1,088
Net Insurance	$	1,365
Relative Value Loss	$	-
Net Cost (Savings)	$	9,864

Social Benefits		
Refueling Time Savings	$	0
Energy Security	$	1
Social Cost of Carbon Emissions	$	-
Crashes, Fatalities, Congestion, Noise	$	-
Lost Fuel Tax Revenue	$	(14)
Net Cost (Savings) Including Social Benefits	$	9,878

Meets 2020 Target	✓
Meets Original 2025 Target	✓
Saves Money (including Social Benefits)	✗
Breakeven cost of Gasoline (per Gallon)	$ -

CAFE Fuel Economy (mpg)	29
On Road mpg Winter (EPA 5-cycle Study)	22
On Road mpg Summer (EPA 5-cycle Study)	24
On Road mpg Typical Driving - Accessories OFF (EPA 5-cycle Study)	25

Acura RLX

MSRP: $ 61,900
Gasoline Counterpart: RLX

Lifetime Economic Costs	
Technology Cost	$ 7,000
Net Fuel Cost (Savings)	$ (978)
Net Taxes & Fees	$ 382
Net Financing	$ 1,072
Net Insurance	$ 1,346
Relative Value Loss	$ -
Net Cost (Savings)	$ 8,823

Social Benefits	
Refueling Time Savings	$ 9
Energy Security	$ 12
Social Cost of Carbon Emissions	$ 866
Crashes, Fatalities, Congestion, Noise	$ (8)
Lost Fuel Tax Revenue	$ (183)
Net Cost (Savings) Including Social Benefits	$ 8,127

Meets 2020 Target	✓
Meets Original 2025 Target	✗
Saves Money (including Social Benefits)	✗
Breakeven cost of Gasoline (per Gallon)	$ 21.31

CAFE Fuel Economy (mpg)	40
On Road mpg Winter (EPA 5-cycle Study)	28
On Road mpg Summer (EPA 5-cycle Study)	30
On Road mpg Typical Driving - Accessories OFF (EPA 5-cycle Study)	31

Honda ACCORD

MSRP: $ 25,470
Gasoline Counterpart: ACCORD

Lifetime Economic Costs		
Technology Cost	$	1,600
Net Fuel Cost (Savings)	$	(1,507)
Net Taxes & Fees	$	87
Net Financing	$	245
Net Insurance	$	308
Relative Value Loss	$	-
Net Cost (Savings)	$	733

Social Benefits		
Refueling Time Savings	$	31
Energy Security	$	44
Social Cost of Carbon Emissions	$	797
Crashes, Fatalities, Congestion, Noise	$	(13)
Lost Fuel Tax Revenue	$	(281)
Net Cost (Savings) Including Social Benefits	$	155

Meets 2020 Target	✓	
Meets Original 2025 Target	✓	
Saves Money (including Social Benefits)	✗	
Breakeven cost of Gasoline (per Gallon)	$	3.51

CAFE Fuel Economy (mpg)		68
On Road mpg Winter (EPA 5-cycle Study)		48
On Road mpg Summer (EPA 5-cycle Study)		48
On Road mpg Typical Driving - Accessories OFF (EPA 5-cycle Study)		45

Honda INSIGHT

MSRP: $ 22,930
Gasoline Counterpart: Civic

Lifetime Economic Costs		
Technology Cost	$	2,380
Net Fuel Cost (Savings)	$	(1,977)
Net Taxes & Fees	$	130
Net Financing	$	365
Net Insurance	$	458
Relative Value Loss	$	-
Net Cost (Savings)	$	1,356

Social Benefits		
Refueling Time Savings	$	46
Energy Security	$	65
Social Cost of Carbon Emissions	$	858
Crashes, Fatalities, Congestion, Noise	$	(17)
Lost Fuel Tax Revenue	$	(369)
Net Cost (Savings) Including Social Benefits	$	772

Meets 2020 Target	✓	
Meets Original 2025 Target	✓	
Saves Money (including Social Benefits)	✗	
Breakeven cost of Gasoline (per Gallon)	$	3.92

CAFE Fuel Economy (mpg)	76
On Road mpg Winter (EPA 5-cycle Study)	55
On Road mpg Summer (EPA 5-cycle Study)	53
On Road mpg Typical Driving - Accessories OFF (EPA 5-cycle Study)	50

Hybrid Electric Vehicles

Honda INSIGHT TOURING

MSRP: $ 28,340
Gasoline Counterpart: Civic

Lifetime Economic Costs	
Technology Cost	$ 7,790
Net Fuel Cost (Savings)	$ (1,464)
Net Taxes & Fees	$ 425
Net Financing	$ 1,193
Net Insurance	$ 1,498
Relative Value Loss	$ -
Net Cost (Savings)	$ 9,443

Social Benefits	
Refueling Time Savings	$ 32
Energy Security	$ 45
Social Cost of Carbon Emissions	$ 779
Crashes, Fatalities, Congestion, Noise	$ (12)
Lost Fuel Tax Revenue	$ (273)
Net Cost (Savings) Including Social Benefits	$ 8,873

Meets 2020 Target	✓
Meets Original 2025 Target	✓
Saves Money (including Social Benefits)	✗
Breakeven cost of Gasoline (per Gallon)	$ 15.94

CAFE Fuel Economy (mpg)	70
On Road mpg Winter (EPA 5-cycle Study)	50
On Road mpg Summer (EPA 5-cycle Study)	49
On Road mpg Typical Driving - Accessories OFF (EPA 5-cycle Study)	46

HYUNDA Ioniq

MSRP: $ 25,880
Gasoline Counterpart: Veloster

Lifetime Economic Costs	
Technology Cost	$ 6,360
Net Fuel Cost (Savings)	$ (2,862)
Net Taxes & Fees	$ 347
Net Financing	$ 974
Net Insurance	$ 1,223
Relative Value Loss	$ -
Net Cost (Savings)	$ 6,043

Social Benefits	
Refueling Time Savings	$ 57
Energy Security	$ 80
Social Cost of Carbon Emissions	$ 1,056
Crashes, Fatalities, Congestion, Noise	$ (24)
Lost Fuel Tax Revenue	$ (534)
Net Cost (Savings) Including Social Benefits	$ 5,408

Meets 2020 Target	✓
Meets Original 2025 Target	✓
Saves Money (including Social Benefits)	✗
Breakeven cost of Gasoline (per Gallon)	$ 6.90

CAFE Fuel Economy (mpg)	74
On Road mpg Winter (EPA 5-cycle Study)	52
On Road mpg Summer (EPA 5-cycle Study)	51
On Road mpg Typical Driving - Accessories OFF (EPA 5-cycle Study)	48

HYUNDA Ioniq Blue

MSRP: $ 23,330
Gasoline Counterpart: Veloster

Lifetime Economic Costs		
Technology Cost	$	3,810
Net Fuel Cost (Savings)	$	(2,949)
Net Taxes & Fees	$	208
Net Financing	$	584
Net Insurance	$	733
Relative Value Loss	$	-
Net Cost (Savings)	$	2,386

Social Benefits		
Refueling Time Savings	$	63
Energy Security	$	88
Social Cost of Carbon Emissions	$	1,058
Crashes, Fatalities, Congestion, Noise	$	(25)
Lost Fuel Tax Revenue	$	(551)
Net Cost (Savings) Including Social Benefits	$	1,752

Meets 2020 Target	✓	
Meets Original 2025 Target	✓	
Saves Money (including Social Benefits)	✗	
Breakeven cost of Gasoline (per Gallon)	$	4.18

CAFE Fuel Economy (mpg)		77
On Road mpg Winter (EPA 5-cycle Study)		55
On Road mpg Summer (EPA 5-cycle Study)		53
On Road mpg Typical Driving - Accessories OFF (EPA 5-cycle Study)		50

Land Rover Discovery Sport MHEV

MSRP: $ 37,800
Gasoline Counterpart: Average non-HEV Differential

Lifetime Economic Costs	
Technology Cost	$ 7,100
Net Fuel Cost (Savings)	$ 3,078
Net Taxes & Fees	$ 388
Net Financing	$ 1,088
Net Insurance	$ 1,365
Relative Value Loss	$ -
Net Cost (Savings)	$ 13,018

Social Benefits	
Refueling Time Savings	$ (18)
Energy Security	$ (26)
Social Cost of Carbon Emissions	$ 144
Crashes, Fatalities, Congestion, Noise	$ 26
Lost Fuel Tax Revenue	$ 575
Net Cost (Savings) Including Social Benefits	$ 12,317

Meets 2020 Target	✓
Meets Original 2025 Target	✗
Saves Money (including Social Benefits)	✗
Breakeven cost of Gasoline (per Gallon)	NA

CAFE Fuel Economy (mpg)	28
On Road mpg Winter (EPA 5-cycle Study)	21
On Road mpg Summer (EPA 5-cycle Study)	23
On Road mpg Typical Driving - Accessories OFF (EPA 5-cycle Study)	24

Land Rover Evoque MHEV

MSRP: $ 42,650
Gasoline Counterpart: Average non-HEV Differential

Lifetime Economic Costs		
Technology Cost	$	7,100
Net Fuel Cost (Savings)	$	1,936
Net Taxes & Fees	$	388
Net Financing	$	1,088
Net Insurance	$	1,365
Relative Value Loss	$	-
Net Cost (Savings)	$	11,876

Social Benefits		
Refueling Time Savings	$	(12)
Energy Security	$	(17)
Social Cost of Carbon Emissions	$	354
Crashes, Fatalities, Congestion, Noise	$	16
Lost Fuel Tax Revenue	$	361
Net Cost (Savings) Including Social Benefits	$	11,174

Meets 2020 Target	✓
Meets Original 2025 Target	✗
Saves Money (including Social Benefits)	✗
Breakeven cost of Gasoline (per Gallon)	NA

CAFE Fuel Economy (mpg)	30
On Road mpg Winter (EPA 5-cycle Study)	22
On Road mpg Summer (EPA 5-cycle Study)	24
On Road mpg Typical Driving - Accessories OFF (EPA 5-cycle Study)	26

Land Rover Range Rover MHEV

MSRP: $ 37,800
Gasoline Counterpart: Average non-HEV Differential

Lifetime Economic Costs	
Technology Cost	$ 7,100
Net Fuel Cost (Savings)	$ 2,871
Net Taxes & Fees	$ 388
Net Financing	$ 1,088
Net Insurance	$ 1,365
Relative Value Loss	$ -
Net Cost (Savings)	$ 12,811

Social Benefits	
Refueling Time Savings	$ (17)
Energy Security	$ (24)
Social Cost of Carbon Emissions	$ 183
Crashes, Fatalities, Congestion, Noise	$ 24
Lost Fuel Tax Revenue	$ 536
Net Cost (Savings) Including Social Benefits	$ 12,110

Meets 2020 Target	✓
Meets Original 2025 Target	✗
Saves Money (including Social Benefits)	✗
Breakeven cost of Gasoline (per Gallon)	NA

CAFE Fuel Economy (mpg)	28
On Road mpg Winter (EPA 5-cycle Study)	21
On Road mpg Summer (EPA 5-cycle Study)	23
On Road mpg Typical Driving - Accessories OFF (EPA 5-cycle Study)	25

KIA Niro

MSRP: $ 23,900
Gasoline Counterpart: Soul

Lifetime Economic Costs	
Technology Cost	$ 6,410
Net Fuel Cost (Savings)	$ (2,387)
Net Taxes & Fees	$ 350
Net Financing	$ 982
Net Insurance	$ 1,233
Relative Value Loss	$ -
Net Cost (Savings)	$ 6,588

Social Benefits	
Refueling Time Savings	$ 38
Energy Security	$ 53
Social Cost of Carbon Emissions	$ 960
Crashes, Fatalities, Congestion, Noise	$ (20)
Lost Fuel Tax Revenue	$ (446)
Net Cost (Savings) Including Social Benefits	$ 6,003

Meets 2020 Target	✓
Meets Original 2025 Target	✓
Saves Money (including Social Benefits)	✗
Breakeven cost of Gasoline (per Gallon)	$ 8.25

CAFE Fuel Economy (mpg)	64
On Road mpg Winter (EPA 5-cycle Study)	45
On Road mpg Summer (EPA 5-cycle Study)	45
On Road mpg Typical Driving - Accessories OFF (EPA 5-cycle Study)	43

KIA Niro FE

MSRP: $ 23,490
Gasoline Counterpart: Soul

Lifetime Economic Costs		
Technology Cost	$	6,000
Net Fuel Cost (Savings)	$	(2,729)
Net Taxes & Fees	$	328
Net Financing	$	919
Net Insurance	$	1,154
Relative Value Loss	$	-
Net Cost (Savings)	$	5,671

Social Benefits		
Refueling Time Savings	$	45
Energy Security	$	63
Social Cost of Carbon Emissions	$	1,010
Crashes, Fatalities, Congestion, Noise	$	(23)
Lost Fuel Tax Revenue	$	(510)
Net Cost (Savings) Including Social Benefits	$	5,086

Meets 2020 Target	✓	
Meets Original 2025 Target	✓	
Saves Money (including Social Benefits)	✗	
Breakeven cost of Gasoline (per Gallon)	$	6.83

CAFE Fuel Economy (mpg)		67
On Road mpg Winter (EPA 5-cycle Study)		48
On Road mpg Summer (EPA 5-cycle Study)		47
On Road mpg Typical Driving - Accessories OFF (EPA 5-cycle Study)		45

KIA Niro Touring

MSRP: $ 32,250
Gasoline Counterpart: Soul

Lifetime Economic Costs	
Technology Cost	$ 14,760
Net Fuel Cost (Savings)	$ (1,767)
Net Taxes & Fees	$ 806
Net Financing	$ 2,261
Net Insurance	$ 2,838
Relative Value Loss	$ -
Net Cost (Savings)	$ 18,898

Social Benefits	
Refueling Time Savings	$ 26
Energy Security	$ 37
Social Cost of Carbon Emissions	$ 869
Crashes, Fatalities, Congestion, Noise	$ (15)
Lost Fuel Tax Revenue	$ (330)
Net Cost (Savings) Including Social Benefits	$ 18,312

Meets 2020 Target	✓
Meets Original 2025 Target	✓
Saves Money (including Social Benefits)	✗
Breakeven cost of Gasoline (per Gallon)	$ 24.80

CAFE Fuel Economy (mpg)	59
On Road mpg Winter (EPA 5-cycle Study)	42
On Road mpg Summer (EPA 5-cycle Study)	42
On Road mpg Typical Driving - Accessories OFF (EPA 5-cycle Study)	41

KIA Optima Hybrid

MSRP: $ 29,310
Gasoline Counterpart: Optima

Lifetime Economic Costs	
Technology Cost	$ 6,120
Net Fuel Cost (Savings)	$ (1,156)
Net Taxes & Fees	$ 334
Net Financing	$ 938
Net Insurance	$ 1,177
Relative Value Loss	$ -
Net Cost (Savings)	$ 7,412

Social Benefits	
Refueling Time Savings	$ 19
Energy Security	$ 26
Social Cost of Carbon Emissions	$ -
Crashes, Fatalities, Congestion, Noise	$ -
Lost Fuel Tax Revenue	$ (216)
Net Cost (Savings) Including Social Benefits	$ 7,583

Meets 2020 Target	✓
Meets Original 2025 Target	✓
Saves Money (including Social Benefits)	✗
Breakeven cost of Gasoline (per Gallon)	$ -

CAFE Fuel Economy (mpg)	58
On Road mpg Winter (EPA 5-cycle Study)	41
On Road mpg Summer (EPA 5-cycle Study)	41
On Road mpg Typical Driving - Accessories OFF (EPA 5-cycle Study)	40

Mercedes-Benz AMG CLS 53 4MATIC+

MSRP: $ 81,200
Gasoline Counterpart: Average non-HEV Differential

Lifetime Economic Costs	
Technology Cost	$ 7,100
Net Fuel Cost (Savings)	$ (1,556)
Net Taxes & Fees	$ 388
Net Financing	$ 1,088
Net Insurance	$ 1,365
Relative Value Loss	$ -
Net Cost (Savings)	$ 8,385

Social Benefits	
Refueling Time Savings	$ 9
Energy Security	$ 13
Social Cost of Carbon Emissions	$ -
Crashes, Fatalities, Congestion, Noise	$ -
Lost Fuel Tax Revenue	$ (290)
Net Cost (Savings) Including Social Benefits	$ 8,654

Meets 2020 Target	✓
Meets Original 2025 Target	✓
Saves Money (including Social Benefits)	✗
Breakeven cost of Gasoline (per Gallon)	$ -

CAFE Fuel Economy (mpg)	31
On Road mpg Winter (EPA 5-cycle Study)	23
On Road mpg Summer (EPA 5-cycle Study)	25
On Road mpg Typical Driving - Accessories OFF (EPA 5-cycle Study)	26

Mercedes-Benz AMG E 53 4MATIC+

MSRP: $ 74,950
Gasoline Counterpart: Average non-HEV Differential

Lifetime Economic Costs	
Technology Cost	$ 7,100
Net Fuel Cost (Savings)	$ (1,836)
Net Taxes & Fees	$ 388
Net Financing	$ 1,088
Net Insurance	$ 1,365
Relative Value Loss	$ -
Net Cost (Savings)	$ 8,105

Social Benefits	
Refueling Time Savings	$ 11
Energy Security	$ 15
Social Cost of Carbon Emissions	$ -
Crashes, Fatalities, Congestion, Noise	$ -
Lost Fuel Tax Revenue	$ (343)
Net Cost (Savings) Including Social Benefits	$ 8,422

Meets 2020 Target	✓
Meets Original 2025 Target	✓
Saves Money (including Social Benefits)	✗
Breakeven cost of Gasoline (per Gallon)	$ -

CAFE Fuel Economy (mpg)	32
On Road mpg Winter (EPA 5-cycle Study)	23
On Road mpg Summer (EPA 5-cycle Study)	25
On Road mpg Typical Driving - Accessories OFF (EPA 5-cycle Study)	27

Mercedes-Benz AMG E 53 4MATIC+ (Convertible)

MSRP: $ 81,650
Gasoline Counterpart: Average non-HEV Differential

Lifetime Economic Costs	
Technology Cost	$ 7,100
Net Fuel Cost (Savings)	$ (1,051)
Net Taxes & Fees	$ 388
Net Financing	$ 1,088
Net Insurance	$ 1,365
Relative Value Loss	$ -
Net Cost (Savings)	$ 8,889

Social Benefits	
Refueling Time Savings	$ 6
Energy Security	$ 8
Social Cost of Carbon Emissions	$ -
Crashes, Fatalities, Congestion, Noise	$ -
Lost Fuel Tax Revenue	$ (196)
Net Cost (Savings) Including Social Benefits	$ 9,072

Meets 2020 Target	✓
Meets Original 2025 Target	✓
Saves Money (including Social Benefits)	✗
Breakeven cost of Gasoline (per Gallon)	$ -

CAFE Fuel Economy (mpg)	30
On Road mpg Winter (EPA 5-cycle Study)	22
On Road mpg Summer (EPA 5-cycle Study)	24
On Road mpg Typical Driving - Accessories OFF (EPA 5-cycle Study)	25

Mercedes-Benz AMG E 53 4MATIC+ (Coupe)

MSRP: $ 74,950
Gasoline Counterpart: Average non-HEV Differential

Lifetime Economic Costs		
Technology Cost	$	7,100
Net Fuel Cost (Savings)	$	(1,491)
Net Taxes & Fees	$	388
Net Financing	$	1,088
Net Insurance	$	1,365
Relative Value Loss	$	-
Net Cost (Savings)	$	8,450

Social Benefits		
Refueling Time Savings	$	8
Energy Security	$	12
Social Cost of Carbon Emissions	$	-
Crashes, Fatalities, Congestion, Noise	$	-
Lost Fuel Tax Revenue	$	(278)
Net Cost (Savings) Including Social Benefits	$	8,708

Meets 2020 Target	✓
Meets Original 2025 Target	✓
Saves Money (including Social Benefits)	✗
Breakeven cost of Gasoline (per Gallon)	$ -

CAFE Fuel Economy (mpg)	31
On Road mpg Winter (EPA 5-cycle Study)	23
On Road mpg Summer (EPA 5-cycle Study)	25
On Road mpg Typical Driving - Accessories OFF (EPA 5-cycle Study)	26

Mercedes-Benz AMG GT 53 4MATIC+

MSRP: $ 127,900
Gasoline Counterpart: Average non-HEV Differential

Lifetime Economic Costs		
Technology Cost	$	7,100
Net Fuel Cost (Savings)	$	(892)
Net Taxes & Fees	$	388
Net Financing	$	1,088
Net Insurance	$	1,365
Relative Value Loss	$	-
Net Cost (Savings)	$	9,049

Social Benefits		
Refueling Time Savings	$	4
Energy Security	$	6
Social Cost of Carbon Emissions	$	-
Crashes, Fatalities, Congestion, Noise	$	-
Lost Fuel Tax Revenue	$	(167)
Net Cost (Savings) Including Social Benefits	$	9,205

Meets 2020 Target	✓
Meets Original 2025 Target	✓
Saves Money (including Social Benefits)	✗
Breakeven cost of Gasoline (per Gallon)	$ -

CAFE Fuel Economy (mpg)	27
On Road mpg Winter (EPA 5-cycle Study)	21
On Road mpg Summer (EPA 5-cycle Study)	22
On Road mpg Typical Driving - Accessories OFF (EPA 5-cycle Study)	24

Mercedes-Benz CLS 450

MSRP: $ 69,950
Gasoline Counterpart: Average non-HEV Differential

Lifetime Economic Costs	
Technology Cost	$ 7,100
Net Fuel Cost (Savings)	$ 2,480
Net Taxes & Fees	$ 388
Net Financing	$ 1,088
Net Insurance	$ 1,365
Relative Value Loss	$ -
Net Cost (Savings)	$ 12,420

Social Benefits	
Refueling Time Savings	$ (24)
Energy Security	$ (34)
Social Cost of Carbon Emissions	$ -
Crashes, Fatalities, Congestion, Noise	$ -
Lost Fuel Tax Revenue	$ 463
Net Cost (Savings) Including Social Benefits	$ 12,016

Meets 2020 Target	✓
Meets Original 2025 Target	✓
Saves Money (including Social Benefits)	✗
Breakeven cost of Gasoline (per Gallon)	$ -

CAFE Fuel Economy (mpg)	36
On Road mpg Winter (EPA 5-cycle Study)	26
On Road mpg Summer (EPA 5-cycle Study)	27
On Road mpg Typical Driving - Accessories OFF (EPA 5-cycle Study)	28

Mercedes-Benz CLS 450 4MATIC

MSRP: $ 72,450
Gasoline Counterpart: Average non-HEV Differential

Lifetime Economic Costs	
Technology Cost	$ 7,100
Net Fuel Cost (Savings)	$ 2,192
Net Taxes & Fees	$ 388
Net Financing	$ 1,088
Net Insurance	$ 1,365
Relative Value Loss	$ -
Net Cost (Savings)	$ 12,133

Social Benefits	
Refueling Time Savings	$ (20)
Energy Security	$ (28)
Social Cost of Carbon Emissions	$ -
Crashes, Fatalities, Congestion, Noise	$ -
Lost Fuel Tax Revenue	$ 409
Net Cost (Savings) Including Social Benefits	$ 11,771

Meets 2020 Target	✓
Meets Original 2025 Target	✓
Saves Money (including Social Benefits)	✗
Breakeven cost of Gasoline (per Gallon)	$ -

CAFE Fuel Economy (mpg)	34
On Road mpg Winter (EPA 5-cycle Study)	25
On Road mpg Summer (EPA 5-cycle Study)	27
On Road mpg Typical Driving - Accessories OFF (EPA 5-cycle Study)	28

Mercedes-Benz GLE 450 4MATIC

MSRP: $ 61,150
Gasoline Counterpart: Average non-HEV Differential

Lifetime Economic Costs		
Technology Cost	$	7,100
Net Fuel Cost (Savings)	$	2,629
Net Taxes & Fees	$	388
Net Financing	$	1,088
Net Insurance	$	1,365
Relative Value Loss	$	-
Net Cost (Savings)	$	12,570

Social Benefits		
Refueling Time Savings	$	(15)
Energy Security	$	(21)
Social Cost of Carbon Emissions	$	-
Crashes, Fatalities, Congestion, Noise	$	-
Lost Fuel Tax Revenue	$	491
Net Cost (Savings) Including Social Benefits	$	12,116

Meets 2020 Target	✓
Meets Original 2025 Target	✓
Saves Money (including Social Benefits)	✗
Breakeven cost of Gasoline (per Gallon)	$ -

CAFE Fuel Economy (mpg)	28
On Road mpg Winter (EPA 5-cycle Study)	21
On Road mpg Summer (EPA 5-cycle Study)	23
On Road mpg Typical Driving - Accessories OFF (EPA 5-cycle Study)	24

Mercedes-Benz GLE 580 4MATIC

MSRP: $ 61,150
Gasoline Counterpart: Average non-HEV Differential

Lifetime Economic Costs		
Technology Cost	$	7,100
Net Fuel Cost (Savings)	$	4,317
Net Taxes & Fees	$	388
Net Financing	$	1,088
Net Insurance	$	1,365
Relative Value Loss	$	-
Net Cost (Savings)	$	14,257

Social Benefits		
Refueling Time Savings	$	(23)
Energy Security	$	(32)
Social Cost of Carbon Emissions	$	-
Crashes, Fatalities, Congestion, Noise	$	-
Lost Fuel Tax Revenue	$	806
Net Cost (Savings) Including Social Benefits	$	13,506

Meets 2020 Target	✓
Meets Original 2025 Target	✓
Saves Money (including Social Benefits)	✗
Breakeven cost of Gasoline (per Gallon)	$ -

CAFE Fuel Economy (mpg)	25
On Road mpg Winter (EPA 5-cycle Study)	19
On Road mpg Summer (EPA 5-cycle Study)	21
On Road mpg Typical Driving - Accessories OFF (EPA 5-cycle Study)	23

Mercedes-Benz GLS 450 4MATIC

MSRP: $ 75,400
Gasoline Counterpart: Average non-HEV Differential

Lifetime Economic Costs		
Technology Cost	$	7,100
Net Fuel Cost (Savings)	$	2,922
Net Taxes & Fees	$	388
Net Financing	$	1,088
Net Insurance	$	1,365
Relative Value Loss	$	-
Net Cost (Savings)	$	12,863

Social Benefits		
Refueling Time Savings	$	(17)
Energy Security	$	(23)
Social Cost of Carbon Emissions	$	-
Crashes, Fatalities, Congestion, Noise	$	-
Lost Fuel Tax Revenue	$	546
Net Cost (Savings) Including Social Benefits	$	12,357

Meets 2020 Target	✓
Meets Original 2025 Target	✓
Saves Money (including Social Benefits)	✗
Breakeven cost of Gasoline (per Gallon)	$ -

CAFE Fuel Economy (mpg)	27
On Road mpg Winter (EPA 5-cycle Study)	21
On Road mpg Summer (EPA 5-cycle Study)	22
On Road mpg Typical Driving - Accessories OFF (EPA 5-cycle Study)	24

Mercedes-Benz GLS 580 4MATIC

MSRP: $ 61,150
Gasoline Counterpart: Average non-HEV Differential

Lifetime Economic Costs	
Technology Cost	$ 7,100
Net Fuel Cost (Savings)	$ 4,965
Net Taxes & Fees	$ 388
Net Financing	$ 1,088
Net Insurance	$ 1,365
Relative Value Loss	$ -
Net Cost (Savings)	$ 14,906

Social Benefits	
Refueling Time Savings	$ (25)
Energy Security	$ (35)
Social Cost of Carbon Emissions	$ -
Crashes, Fatalities, Congestion, Noise	$ -
Lost Fuel Tax Revenue	$ 927
Net Cost (Savings) Including Social Benefits	$ 14,039

Meets 2020 Target	✓
Meets Original 2025 Target	✓
Saves Money (including Social Benefits)	✗
Breakeven cost of Gasoline (per Gallon)	$ -

CAFE Fuel Economy (mpg)	24
On Road mpg Winter (EPA 5-cycle Study)	19
On Road mpg Summer (EPA 5-cycle Study)	20
On Road mpg Typical Driving - Accessories OFF (EPA 5-cycle Study)	22

LEXUS ES 300h

MSRP: $ 41,760
Gasoline Counterpart: ES 350

Lifetime Economic Costs		
Technology Cost	$	1,860
Net Fuel Cost (Savings)	$	(3,396)
Net Taxes & Fees	$	102
Net Financing	$	285
Net Insurance	$	358
Relative Value Loss	$	-
Net Cost (Savings)	$	(792)

Social Benefits		
Refueling Time Savings	$	47
Energy Security	$	67
Social Cost of Carbon Emissions	$	1,234
Crashes, Fatalities, Congestion, Noise	$	(29)
Lost Fuel Tax Revenue	$	(634)
Net Cost (Savings) Including Social Benefits	$	(1,477)

Meets 2020 Target	✓	
Meets Original 2025 Target	✓	
Saves Money (including Social Benefits)	✓	
Breakeven cost of Gasoline (per Gallon)	$	2.01

CAFE Fuel Economy (mpg)	60
On Road mpg Winter (EPA 5-cycle Study)	42
On Road mpg Summer (EPA 5-cycle Study)	43
On Road mpg Typical Driving - Accessories OFF (EPA 5-cycle Study)	41

LEXUS LC 500h

MSRP: $ 97,460
Gasoline Counterpart: LC 500

Lifetime Economic Costs		
Technology Cost	$	7,100
Net Fuel Cost (Savings)	$	(4,121)
Net Taxes & Fees	$	388
Net Financing	$	1,088
Net Insurance	$	1,365
Relative Value Loss	$	-
Net Cost (Savings)	$	5,820

Social Benefits		
Refueling Time Savings	$	30
Energy Security	$	43
Social Cost of Carbon Emissions	$	-
Crashes, Fatalities, Congestion, Noise	$	-
Lost Fuel Tax Revenue	$	(769)
Net Cost (Savings) Including Social Benefits	$	6,516

Meets 2020 Target	✓
Meets Original 2025 Target	✓
Saves Money (including Social Benefits)	✗
Breakeven cost of Gasoline (per Gallon)	$ -

CAFE Fuel Economy (mpg)	41
On Road mpg Winter (EPA 5-cycle Study)	29
On Road mpg Summer (EPA 5-cycle Study)	31
On Road mpg Typical Driving - Accessories OFF (EPA 5-cycle Study)	31

LEXUS LS 500h

MSRP: $ 79,960
Gasoline Counterpart: LS 500

Lifetime Economic Costs	
Technology Cost	$ 5,510
Net Fuel Cost (Savings)	$ (768)
Net Taxes & Fees	$ 301
Net Financing	$ 844
Net Insurance	$ 1,060
Relative Value Loss	$ -
Net Cost (Savings)	$ 6,946

Social Benefits	
Refueling Time Savings	$ 6
Energy Security	$ 9
Social Cost of Carbon Emissions	$ 834
Crashes, Fatalities, Congestion, Noise	$ (6)
Lost Fuel Tax Revenue	$ (143)
Net Cost (Savings) Including Social Benefits	$ 6,247

Meets 2020 Target	✓
Meets Original 2025 Target	✗
Saves Money (including Social Benefits)	✗
Breakeven cost of Gasoline (per Gallon)	$ 21.35

CAFE Fuel Economy (mpg)	38
On Road mpg Winter (EPA 5-cycle Study)	27
On Road mpg Summer (EPA 5-cycle Study)	29
On Road mpg Typical Driving - Accessories OFF (EPA 5-cycle Study)	30

LEXUS LS 500h AWD

MSRP: $ 83,180
Gasoline Counterpart: LS 500 AWD

Lifetime Economic Costs		
Technology Cost	$	4,510
Net Fuel Cost (Savings)	$	(687)
Net Taxes & Fees	$	246
Net Financing	$	691
Net Insurance	$	867
Relative Value Loss	$	-
Net Cost (Savings)	$	5,628

Social Benefits		
Refueling Time Savings	$	5
Energy Security	$	7
Social Cost of Carbon Emissions	$	842
Crashes, Fatalities, Congestion, Noise	$	(6)
Lost Fuel Tax Revenue	$	(128)
Net Cost (Savings) Including Social Benefits	$	4,907

Meets 2020 Target	✓	
Meets Original 2025 Target	✗	
Saves Money (including Social Benefits)	✗	
Breakeven cost of Gasoline (per Gallon)	$	19.58

CAFE Fuel Economy (mpg)	35
On Road mpg Winter (EPA 5-cycle Study)	25
On Road mpg Summer (EPA 5-cycle Study)	27
On Road mpg Typical Driving - Accessories OFF (EPA 5-cycle Study)	28

LEXUS NX 300h AWD

MSRP: $ 39,270
Gasoline Counterpart: NX 300 AWD

Lifetime Economic Costs		
Technology Cost	$	1,000
Net Fuel Cost (Savings)	$	(1,318)
Net Taxes & Fees	$	55
Net Financing	$	153
Net Insurance	$	192
Relative Value Loss	$	-
Net Cost (Savings)	$	82

Social Benefits		
Refueling Time Savings	$	12
Energy Security	$	17
Social Cost of Carbon Emissions	$	886
Crashes, Fatalities, Congestion, Noise	$	(11)
Lost Fuel Tax Revenue	$	(246)
Net Cost (Savings) Including Social Benefits	$	(576)

Meets 2020 Target	✓	
Meets Original 2025 Target	✗	
Saves Money (including Social Benefits)	✓	
Breakeven cost of Gasoline (per Gallon)	$	2.62

CAFE Fuel Economy (mpg)	44
On Road mpg Winter (EPA 5-cycle Study)	31
On Road mpg Summer (EPA 5-cycle Study)	32
On Road mpg Typical Driving - Accessories OFF (EPA 5-cycle Study)	33

LEXUS RX 450h AWD

MSRP: $ 46,800
Gasoline Counterpart: RX 350L AWD

Lifetime Economic Costs		
Technology Cost	$	1,250
Net Fuel Cost (Savings)	$	(2,534)
Net Taxes & Fees	$	68
Net Financing	$	192
Net Insurance	$	240
Relative Value Loss	$	-
Net Cost (Savings)	$	(784)

Social Benefits		
Refueling Time Savings	$	19
Energy Security	$	27
Social Cost of Carbon Emissions	$	1,138
Crashes, Fatalities, Congestion, Noise	$	(21)
Lost Fuel Tax Revenue	$	(473)
Net Cost (Savings) Including Social Benefits	$	(1,475)

Meets 2020 Target	✓	
Meets Original 2025 Target	✗	
Saves Money (including Social Benefits)	✓	
Breakeven cost of Gasoline (per Gallon)	$	1.85

CAFE Fuel Economy (mpg)	41
On Road mpg Winter (EPA 5-cycle Study)	29
On Road mpg Summer (EPA 5-cycle Study)	31
On Road mpg Typical Driving - Accessories OFF (EPA 5-cycle Study)	31

LEXUS RX 450hL AWD

MSRP: $ 50,460
Gasoline Counterpart: RX 350L AWD

Lifetime Economic Costs		
Technology Cost	$	4,910
Net Fuel Cost (Savings)	$	(2,534)
Net Taxes & Fees	$	268
Net Financing	$	752
Net Insurance	$	944
Relative Value Loss	$	-
Net Cost (Savings)	$	4,340

Social Benefits		
Refueling Time Savings	$	19
Energy Security	$	27
Social Cost of Carbon Emissions	$	1,138
Crashes, Fatalities, Congestion, Noise	$	(21)
Lost Fuel Tax Revenue	$	(473)
Net Cost (Savings) Including Social Benefits	$	3,650

Meets 2020 Target	✓
Meets Original 2025 Target	✗
Saves Money (including Social Benefits)	✗
Breakeven cost of Gasoline (per Gallon)	$ 6.07

CAFE Fuel Economy (mpg)	41
On Road mpg Winter (EPA 5-cycle Study)	29
On Road mpg Summer (EPA 5-cycle Study)	31
On Road mpg Typical Driving - Accessories OFF (EPA 5-cycle Study)	31

LEXUS UX 250h

MSRP: $ 34,350
Gasoline Counterpart: UX

Lifetime Economic Costs		
Technology Cost	$	2,050
Net Fuel Cost (Savings)	$	(756)
Net Taxes & Fees	$	112
Net Financing	$	314
Net Insurance	$	394
Relative Value Loss	$	-
Net Cost (Savings)	$	2,114

Social Benefits		
Refueling Time Savings	$	12
Energy Security	$	17
Social Cost of Carbon Emissions	$	678
Crashes, Fatalities, Congestion, Noise	$	(6)
Lost Fuel Tax Revenue	$	(141)
Net Cost (Savings) Including Social Benefits	$	1,554

Meets 2020 Target	✓	
Meets Original 2025 Target	✓	
Saves Money (including Social Benefits)	✗	
Breakeven cost of Gasoline (per Gallon)	$	8.32

CAFE Fuel Economy (mpg)		59
On Road mpg Winter (EPA 5-cycle Study)		41
On Road mpg Summer (EPA 5-cycle Study)		42
On Road mpg Typical Driving - Accessories OFF (EPA 5-cycle Study)		40

LEXUS UX 250h AWD

MSRP: $ 34,350
Gasoline Counterpart: UX

Lifetime Economic Costs		
Technology Cost	$	2,050
Net Fuel Cost (Savings)	$	(71)
Net Taxes & Fees	$	112
Net Financing	$	314
Net Insurance	$	394
Relative Value Loss	$	-
Net Cost (Savings)	$	2,799

Social Benefits		
Refueling Time Savings	$	1
Energy Security	$	2
Social Cost of Carbon Emissions	$	567
Crashes, Fatalities, Congestion, Noise	$	(1)
Lost Fuel Tax Revenue	$	(13)
Net Cost (Savings) Including Social Benefits	$	2,243

Meets 2020 Target	✓	
Meets Original 2025 Target	✓	
Saves Money (including Social Benefits)	✗	
Breakeven cost of Gasoline (per Gallon)	$	85.02

CAFE Fuel Economy (mpg)		55
On Road mpg Winter (EPA 5-cycle Study)		39
On Road mpg Summer (EPA 5-cycle Study)		40
On Road mpg Typical Driving - Accessories OFF (EPA 5-cycle Study)		39

TOYOTA AVALON HYBRID

MSRP: $ 43,150
Gasoline Counterpart: AVALON

Lifetime Economic Costs	
Technology Cost	$ 1,050
Net Fuel Cost (Savings)	$ (3,522)
Net Taxes & Fees	$ 57
Net Financing	$ 161
Net Insurance	$ 202
Relative Value Loss	$ -
Net Cost (Savings)	$ (2,052)

Social Benefits	
Refueling Time Savings	$ 47
Energy Security	$ 67
Social Cost of Carbon Emissions	$ 1,266
Crashes, Fatalities, Congestion, Noise	$ (30)
Lost Fuel Tax Revenue	$ (658)
Net Cost (Savings) Including Social Benefits	$ (2,745)

Meets 2020 Target	✓
Meets Original 2025 Target	✓
Saves Money (including Social Benefits)	✓
Breakeven cost of Gasoline (per Gallon)	$ 1.28

CAFE Fuel Economy (mpg)	59
On Road mpg Winter (EPA 5-cycle Study)	42
On Road mpg Summer (EPA 5-cycle Study)	42
On Road mpg Typical Driving - Accessories OFF (EPA 5-cycle Study)	41

TOYOTA AVALON HYBRID XLE

MSRP: $ 36,850
Gasoline Counterpart: AVALON XLE

Lifetime Economic Costs		
Technology Cost	$	1,050
Net Fuel Cost (Savings)	$	(3,306)
Net Taxes & Fees	$	57
Net Financing	$	161
Net Insurance	$	202
Relative Value Loss	$	-
Net Cost (Savings)	$	(1,836)

Social Benefits		
Refueling Time Savings	$	47
Energy Security	$	67
Social Cost of Carbon Emissions	$	1,211
Crashes, Fatalities, Congestion, Noise	$	(28)
Lost Fuel Tax Revenue	$	(617)
Net Cost (Savings) Including Social Benefits	$	(2,516)

Meets 2020 Target	✓	
Meets Original 2025 Target	✓	
Saves Money (including Social Benefits)	✓	
Breakeven cost of Gasoline (per Gallon)	$	1.34

CAFE Fuel Economy (mpg)	61
On Road mpg Winter (EPA 5-cycle Study)	43
On Road mpg Summer (EPA 5-cycle Study)	43
On Road mpg Typical Driving - Accessories OFF (EPA 5-cycle Study)	42

TOYOTA CAMRY HYBRID LE

MSRP: $ 28,250
Gasoline Counterpart: CAMRY LE

Lifetime Economic Costs		
Technology Cost	$	3,410
Net Fuel Cost (Savings)	$	(1,660)
Net Taxes & Fees	$	186
Net Financing	$	522
Net Insurance	$	656
Relative Value Loss	$	-
Net Cost (Savings)	$	3,114

Social Benefits		
Refueling Time Savings	$	36
Energy Security	$	51
Social Cost of Carbon Emissions	$	813
Crashes, Fatalities, Congestion, Noise	$	(14)
Lost Fuel Tax Revenue	$	(310)
Net Cost (Savings) Including Social Benefits	$	2,538

Meets 2020 Target	✓	
Meets Original 2025 Target	✓	
Saves Money (including Social Benefits)	✗	
Breakeven cost of Gasoline (per Gallon)	$	6.41

CAFE Fuel Economy (mpg)		72
On Road mpg Winter (EPA 5-cycle Study)		51
On Road mpg Summer (EPA 5-cycle Study)		50
On Road mpg Typical Driving - Accessories OFF (EPA 5-cycle Study)		47

Hybrid Electric Vehicles

TOYOTA CAMRY HYBRID XLE/SE

MSRP: $ 32,550
Gasoline Counterpart: CAMRY XLE/SE

Lifetime Economic Costs		
Technology Cost	$	3,200
Net Fuel Cost (Savings)	$	(1,400)
Net Taxes & Fees	$	175
Net Financing	$	490
Net Insurance	$	615
Relative Value Loss	$	-
Net Cost (Savings)	$	3,081

Social Benefits		
Refueling Time Savings	$	25
Energy Security	$	36
Social Cost of Carbon Emissions	$	803
Crashes, Fatalities, Congestion, Noise	$	(12)
Lost Fuel Tax Revenue	$	(261)
Net Cost (Savings) Including Social Benefits	$	2,490

Meets 2020 Target	✓	
Meets Original 2025 Target	✓	
Saves Money (including Social Benefits)	✗	
Breakeven cost of Gasoline (per Gallon)	$	7.08

CAFE Fuel Economy (mpg)	63
On Road mpg Winter (EPA 5-cycle Study)	44
On Road mpg Summer (EPA 5-cycle Study)	44
On Road mpg Typical Driving - Accessories OFF (EPA 5-cycle Study)	43

TOYOTA COROLLA HYBRID

MSRP: $ 23,100
Gasoline Counterpart: COROLLA

Lifetime Economic Costs	
Technology Cost	$ 3,500
Net Fuel Cost (Savings)	$ (2,004)
Net Taxes & Fees	$ 191
Net Financing	$ 536
Net Insurance	$ 673
Relative Value Loss	$ -
Net Cost (Savings)	$ 2,896

Social Benefits	
Refueling Time Savings	$ 47
Energy Security	$ 67
Social Cost of Carbon Emissions	$ 861
Crashes, Fatalities, Congestion, Noise	$ (17)
Lost Fuel Tax Revenue	$ (374)
Net Cost (Savings) Including Social Benefits	$ 2,313

Meets 2020 Target	✓
Meets Original 2025 Target	✓
Saves Money (including Social Benefits)	✗
Breakeven cost of Gasoline (per Gallon)	$ 5.51

CAFE Fuel Economy (mpg)	77
On Road mpg Winter (EPA 5-cycle Study)	55
On Road mpg Summer (EPA 5-cycle Study)	53
On Road mpg Typical Driving - Accessories OFF (EPA 5-cycle Study)	50

TOYOTA PRIUS

MSRP: $ 24,200
Gasoline Counterpart: COROLLA

Lifetime Economic Costs		
Technology Cost	$	4,600
Net Fuel Cost (Savings)	$	(1,701)
Net Taxes & Fees	$	251
Net Financing	$	705
Net Insurance	$	885
Relative Value Loss	$	-
Net Cost (Savings)	$	4,740

Social Benefits		
Refueling Time Savings	$	38
Energy Security	$	54
Social Cost of Carbon Emissions	$	814
Crashes, Fatalities, Congestion, Noise	$	(14)
Lost Fuel Tax Revenue	$	(318)
Net Cost (Savings) Including Social Benefits	$	4,165

Meets 2020 Target	✓	
Meets Original 2025 Target	✓	
Saves Money (including Social Benefits)	✗	
Breakeven cost of Gasoline (per Gallon)	$	8.31

CAFE Fuel Economy (mpg)	73
On Road mpg Winter (EPA 5-cycle Study)	52
On Road mpg Summer (EPA 5-cycle Study)	51
On Road mpg Typical Driving - Accessories OFF (EPA 5-cycle Study)	48

TOYOTA PRIUS AWD

MSRP: $ 26,800
Gasoline Counterpart: COROLLA

Lifetime Economic Costs		
Technology Cost	$	7,200
Net Fuel Cost (Savings)	$	(1,658)
Net Taxes & Fees	$	393
Net Financing	$	1,103
Net Insurance	$	1,385
Relative Value Loss	$	-
Net Cost (Savings)	$	8,423

Social Benefits		
Refueling Time Savings	$	37
Energy Security	$	52
Social Cost of Carbon Emissions	$	807
Crashes, Fatalities, Congestion, Noise	$	(14)
Lost Fuel Tax Revenue	$	(310)
Net Cost (Savings) Including Social Benefits	$	7,849

Meets 2020 Target	✓	
Meets Original 2025 Target	✓	
Saves Money (including Social Benefits)	✗	
Breakeven cost of Gasoline (per Gallon)	$	13.09

CAFE Fuel Economy (mpg)		73
On Road mpg Winter (EPA 5-cycle Study)		52
On Road mpg Summer (EPA 5-cycle Study)		50
On Road mpg Typical Driving - Accessories OFF (EPA 5-cycle Study)		48

TOYOTA PRIUS Eco

MSRP: $ 24,200
Gasoline Counterpart: COROLLA

Lifetime Economic Costs		
Technology Cost	$	4,600
Net Fuel Cost (Savings)	$	(2,271)
Net Taxes & Fees	$	251
Net Financing	$	705
Net Insurance	$	885
Relative Value Loss	$	-
Net Cost (Savings)	$	4,170

Social Benefits		
Refueling Time Savings	$	56
Energy Security	$	79
Social Cost of Carbon Emissions	$	902
Crashes, Fatalities, Congestion, Noise	$	(19)
Lost Fuel Tax Revenue	$	(424)
Net Cost (Savings) Including Social Benefits	$	3,576

Meets 2020 Target	✓	
Meets Original 2025 Target	✓	
Saves Money (including Social Benefits)	✗	
Breakeven cost of Gasoline (per Gallon)	$	6.32

CAFE Fuel Economy (mpg)		81
On Road mpg Winter (EPA 5-cycle Study)		58
On Road mpg Summer (EPA 5-cycle Study)		55
On Road mpg Typical Driving - Accessories OFF (EPA 5-cycle Study)		52

TOYOTA RAV4 HYBRID AWD

MSRP: $ 28,100
Gasoline Counterpart: RAV4 AWD

Lifetime Economic Costs	
Technology Cost	$ 850
Net Fuel Cost (Savings)	$ (1,416)
Net Taxes & Fees	$ 46
Net Financing	$ 130
Net Insurance	$ 163
Relative Value Loss	$ -
Net Cost (Savings)	$ (226)

Social Benefits	
Refueling Time Savings	$ 19
Energy Security	$ 27
Social Cost of Carbon Emissions	$ 823
Crashes, Fatalities, Congestion, Noise	$ (12)
Lost Fuel Tax Revenue	$ (264)
Net Cost (Savings) Including Social Benefits	$ (819)

Meets 2020 Target	✓
Meets Original 2025 Target	✓
Saves Money (including Social Benefits)	✓
Breakeven cost of Gasoline (per Gallon)	$ 2.16

CAFE Fuel Economy (mpg)	56
On Road mpg Winter (EPA 5-cycle Study)	39
On Road mpg Summer (EPA 5-cycle Study)	40
On Road mpg Typical Driving - Accessories OFF (EPA 5-cycle Study)	39

Audi A6 Allroad

MSRP: $ 45,700
Gasoline Counterpart: Average non-HEV Differential

Lifetime Economic Costs		
Technology Cost	$	7,100
Net Fuel Cost (Savings)	$	3,680
Net Taxes & Fees	$	388
Net Financing	$	1,088
Net Insurance	$	1,365
Relative Value Loss	$	-
Net Cost (Savings)	$	13,620

Social Benefits		
Refueling Time Savings	$	(27)
Energy Security	$	(38)
Social Cost of Carbon Emissions	$	-
Crashes, Fatalities, Congestion, Noise	$	-
Lost Fuel Tax Revenue	$	687
Net Cost (Savings) Including Social Benefits	$	12,999

Meets 2020 Target	✓
Meets Original 2025 Target	✓
Saves Money (including Social Benefits)	✗
Breakeven cost of Gasoline (per Gallon)	$ -

CAFE Fuel Economy (mpg)	31
On Road mpg Winter (EPA 5-cycle Study)	23
On Road mpg Summer (EPA 5-cycle Study)	24
On Road mpg Typical Driving - Accessories OFF (EPA 5-cycle Study)	26

Audi A6 quattro

MSRP: $ 69,000
Gasoline Counterpart: Average non-HEV Differential

Lifetime Economic Costs		
Technology Cost	$	7,100
Net Fuel Cost (Savings)	$	1,493
Net Taxes & Fees	$	388
Net Financing	$	1,088
Net Insurance	$	1,365
Relative Value Loss	$	-
Net Cost (Savings)	$	11,434

Social Benefits		
Refueling Time Savings	$	(14)
Energy Security	$	(20)
Social Cost of Carbon Emissions	$	-
Crashes, Fatalities, Congestion, Noise	$	-
Lost Fuel Tax Revenue	$	279
Net Cost (Savings) Including Social Benefits	$	11,189

Meets 2020 Target	✓
Meets Original 2025 Target	✓
Saves Money (including Social Benefits)	✗
Breakeven cost of Gasoline (per Gallon)	$ -

CAFE Fuel Economy (mpg)	36
On Road mpg Winter (EPA 5-cycle Study)	26
On Road mpg Summer (EPA 5-cycle Study)	28
On Road mpg Typical Driving - Accessories OFF (EPA 5-cycle Study)	29

Audi A6 quattro

MSRP: $ 69,000
Gasoline Counterpart: Average non-HEV Differential

Lifetime Economic Costs		
Technology Cost	$	7,100
Net Fuel Cost (Savings)	$	2,588
Net Taxes & Fees	$	388
Net Financing	$	1,088
Net Insurance	$	1,365
Relative Value Loss	$	-
Net Cost (Savings)	$	12,529

Social Benefits		
Refueling Time Savings	$	(22)
Energy Security	$	(31)
Social Cost of Carbon Emissions	$	-
Crashes, Fatalities, Congestion, Noise	$	-
Lost Fuel Tax Revenue	$	483
Net Cost (Savings) Including Social Benefits	$	12,099

Meets 2020 Target	✓
Meets Original 2025 Target	✓
Saves Money (including Social Benefits)	✗
Breakeven cost of Gasoline (per Gallon)	$ -

CAFE Fuel Economy (mpg)	33
On Road mpg Winter (EPA 5-cycle Study)	24
On Road mpg Summer (EPA 5-cycle Study)	26
On Road mpg Typical Driving - Accessories OFF (EPA 5-cycle Study)	27

Audi A7 quattro

MSRP: $ 54,900
Gasoline Counterpart: Average non-HEV Differential

Lifetime Economic Costs	
Technology Cost	$ 7,100
Net Fuel Cost (Savings)	$ 2,588
Net Taxes & Fees	$ 388
Net Financing	$ 1,088
Net Insurance	$ 1,365
Relative Value Loss	$ -
Net Cost (Savings)	$ 12,529

Social Benefits	
Refueling Time Savings	$ (22)
Energy Security	$ (31)
Social Cost of Carbon Emissions	$ -
Crashes, Fatalities, Congestion, Noise	$ -
Lost Fuel Tax Revenue	$ 483
Net Cost (Savings) Including Social Benefits	$ 12,099

Meets 2020 Target	✓
Meets Original 2025 Target	✓
Saves Money (including Social Benefits)	✗
Breakeven cost of Gasoline (per Gallon)	$ -

CAFE Fuel Economy (mpg)	33
On Road mpg Winter (EPA 5-cycle Study)	24
On Road mpg Summer (EPA 5-cycle Study)	26
On Road mpg Typical Driving - Accessories OFF (EPA 5-cycle Study)	27

Audi A8L

MSRP: $ 85,200
Gasoline Counterpart: Average non-HEV Differential

Lifetime Economic Costs		
Technology Cost	$	7,100
Net Fuel Cost (Savings)	$	4,339
Net Taxes & Fees	$	388
Net Financing	$	1,088
Net Insurance	$	1,365
Relative Value Loss	$	-
Net Cost (Savings)	$	14,280

Social Benefits		
Refueling Time Savings	$	(33)
Energy Security	$	(46)
Social Cost of Carbon Emissions	$	-
Crashes, Fatalities, Congestion, Noise	$	-
Lost Fuel Tax Revenue	$	810
Net Cost (Savings) Including Social Benefits	$	13,549

Meets 2020 Target	✓
Meets Original 2025 Target	✓
Saves Money (including Social Benefits)	✗
Breakeven cost of Gasoline (per Gallon)	$ -

CAFE Fuel Economy (mpg)	28
On Road mpg Winter (EPA 5-cycle Study)	21
On Road mpg Summer (EPA 5-cycle Study)	23
On Road mpg Typical Driving - Accessories OFF (EPA 5-cycle Study)	25

Audi A8L

MSRP: $ 96,800
Gasoline Counterpart: Average non-HEV Differential

Lifetime Economic Costs		
Technology Cost	$	7,100
Net Fuel Cost (Savings)	$	6,434
Net Taxes & Fees	$	388
Net Financing	$	1,088
Net Insurance	$	1,365
Relative Value Loss	$	-
Net Cost (Savings)	$	16,375

Social Benefits		
Refueling Time Savings	$	(43)
Energy Security	$	(61)
Social Cost of Carbon Emissions	$	-
Crashes, Fatalities, Congestion, Noise	$	-
Lost Fuel Tax Revenue	$	1,201
Net Cost (Savings) Including Social Benefits	$	15,277

Meets 2020 Target	✓
Meets Original 2025 Target	✓
Saves Money (including Social Benefits)	✗
Breakeven cost of Gasoline (per Gallon)	$ -

CAFE Fuel Economy (mpg)	24
On Road mpg Winter (EPA 5-cycle Study)	19
On Road mpg Summer (EPA 5-cycle Study)	20
On Road mpg Typical Driving - Accessories OFF (EPA 5-cycle Study)	22

Audi Q7

MSRP: $ 53,550
Gasoline Counterpart: Average non-HEV Differential

Lifetime Economic Costs	
Technology Cost	$ 7,100
Net Fuel Cost (Savings)	$ 5,282
Net Taxes & Fees	$ 388
Net Financing	$ 1,088
Net Insurance	$ 1,365
Relative Value Loss	$ -
Net Cost (Savings)	$ 15,223

Social Benefits	
Refueling Time Savings	$ (30)
Energy Security	$ (43)
Social Cost of Carbon Emissions	$ -
Crashes, Fatalities, Congestion, Noise	$ -
Lost Fuel Tax Revenue	$ 986
Net Cost (Savings) Including Social Benefits	$ 14,310

Meets 2020 Target	✓
Meets Original 2025 Target	✓
Saves Money (including Social Benefits)	✗
Breakeven cost of Gasoline (per Gallon)	$ -

CAFE Fuel Economy (mpg)	25
On Road mpg Winter (EPA 5-cycle Study)	19
On Road mpg Summer (EPA 5-cycle Study)	21
On Road mpg Typical Driving - Accessories OFF (EPA 5-cycle Study)	23

Audi Q8

MSRP: $ 68,200
Gasoline Counterpart: Average non-HEV Differential

Lifetime Economic Costs		
Technology Cost	$	7,100
Net Fuel Cost (Savings)	$	5,282
Net Taxes & Fees	$	388
Net Financing	$	1,088
Net Insurance	$	1,365
Relative Value Loss	$	-
Net Cost (Savings)	$	15,223

Social Benefits		
Refueling Time Savings	$	(30)
Energy Security	$	(43)
Social Cost of Carbon Emissions	$	-
Crashes, Fatalities, Congestion, Noise	$	-
Lost Fuel Tax Revenue	$	986
Net Cost (Savings) Including Social Benefits	$	14,310

Meets 2020 Target	✓
Meets Original 2025 Target	✓
Saves Money (including Social Benefits)	✗
Breakeven cost of Gasoline (per Gallon)	$ -

CAFE Fuel Economy (mpg)	25
On Road mpg Winter (EPA 5-cycle Study)	19
On Road mpg Summer (EPA 5-cycle Study)	21
On Road mpg Typical Driving - Accessories OFF (EPA 5-cycle Study)	23

Audi S6

MSRP: $ 73,900
Gasoline Counterpart: Average non-HEV Differential

Lifetime Economic Costs		
Technology Cost	$	7,100
Net Fuel Cost (Savings)	$	2,279
Net Taxes & Fees	$	388
Net Financing	$	1,088
Net Insurance	$	1,365
Relative Value Loss	$	-
Net Cost (Savings)	$	12,219

Social Benefits		
Refueling Time Savings	$	(15)
Energy Security	$	(21)
Social Cost of Carbon Emissions	$	-
Crashes, Fatalities, Congestion, Noise	$	-
Lost Fuel Tax Revenue	$	425
Net Cost (Savings) Including Social Benefits	$	11,830

Meets 2020 Target	✓
Meets Original 2025 Target	✓
Saves Money (including Social Benefits)	✗
Breakeven cost of Gasoline (per Gallon)	$ -

CAFE Fuel Economy (mpg)	29
On Road mpg Winter (EPA 5-cycle Study)	22
On Road mpg Summer (EPA 5-cycle Study)	23
On Road mpg Typical Driving - Accessories OFF (EPA 5-cycle Study)	25

Audi S7

MSRP: $ 83,900
Gasoline Counterpart: Average non-HEV Differential

Lifetime Economic Costs		
Technology Cost	$	7,100
Net Fuel Cost (Savings)	$	2,279
Net Taxes & Fees	$	388
Net Financing	$	1,088
Net Insurance	$	1,365
Relative Value Loss	$	-
Net Cost (Savings)	$	12,219

Social Benefits		
Refueling Time Savings	$	(15)
Energy Security	$	(21)
Social Cost of Carbon Emissions	$	-
Crashes, Fatalities, Congestion, Noise	$	-
Lost Fuel Tax Revenue	$	425
Net Cost (Savings) Including Social Benefits	$	11,830

Meets 2020 Target	✓	
Meets Original 2025 Target	✓	
Saves Money (including Social Benefits)	✗	
Breakeven cost of Gasoline (per Gallon)	$	-

CAFE Fuel Economy (mpg)		29
On Road mpg Winter (EPA 5-cycle Study)		22
On Road mpg Summer (EPA 5-cycle Study)		23
On Road mpg Typical Driving - Accessories OFF (EPA 5-cycle Study)		25

Audi S8

MSRP: $ 129,500
Gasoline Counterpart: Average non-HEV Differential

Lifetime Economic Costs	
Technology Cost	$ 7,100
Net Fuel Cost (Savings)	$ 6,213
Net Taxes & Fees	$ 388
Net Financing	$ 1,088
Net Insurance	$ 1,365
Relative Value Loss	$ -
Net Cost (Savings)	$ 16,154

Social Benefits	
Refueling Time Savings	$ (32)
Energy Security	$ (46)
Social Cost of Carbon Emissions	$ -
Crashes, Fatalities, Congestion, Noise	$ -
Lost Fuel Tax Revenue	$ 1,160
Net Cost (Savings) Including Social Benefits	$ 15,072

Meets 2020 Target	✓
Meets Original 2025 Target	✓
Saves Money (including Social Benefits)	✗
Breakeven cost of Gasoline (per Gallon)	$ -

CAFE Fuel Economy (mpg)	21
On Road mpg Winter (EPA 5-cycle Study)	17
On Road mpg Summer (EPA 5-cycle Study)	19
On Road mpg Typical Driving - Accessories OFF (EPA 5-cycle Study)	21

DIESEL VEHICLES

The 2020 Model Year EPA/NHTSA Fuel Economy Guide contains information on sixteen advanced diesel vehicles. **Eleven of these vehicles met their 2020 model year fuel economy targets** and twenty-five percent (four) met their 2025 model year fuel economy targets.

None of the diesel-powered vehicles saved the customer money using the methodology employed by the National Highway Traffic Safety Administration and EPA for calculating the cost and benefits of technology.

The average cost premium for a diesel-powered vehicle was $13,200.

Jeep Wrangler Unlimited 4x4

MSRP: $ 35,795
Gasoline Counterpart: Wrangler Unlimited 4x4

Lifetime Economic Costs		
Technology Cost	$	4,000
Net Fuel Cost (Savings)	$	2,980
Net Taxes & Fees	$	218
Net Financing	$	613
Net Insurance	$	769
Relative Value Loss	$	-
Net Cost (Savings)	$	8,580

Social Benefits		
Refueling Time Savings	$	4
Energy Security	$	6
Social Cost of Carbon Emissions	$	433
Crashes, Fatalities, Congestion, Noise	$	25
Lost Fuel Tax Revenue	$	(89)
Net Cost (Savings) Including Social Benefits	$	8,201

Meets 2020 Target	✓	
Meets Original 2025 Target	✗	
Saves Money (including Social Benefits)	✗	
Breakeven cost of Gasoline (per Gallon)	$	3.95

CAFE Fuel Economy (mpg)	33
On Road mpg Winter (EPA 5-cycle Study)	28
On Road mpg Summer (EPA 5-cycle Study)	27
On Road mpg Typical Driving - Accessories OFF (EPA 5-cycle Study)	32

Assumes Diesel and Electricity Prices do not change in relation to gasoline

RAM 1500 4X2

MSRP: $ 33,287
Gasoline Counterpart: 1500 4X2

Lifetime Economic Costs		
Technology Cost	$	2,645
Net Fuel Cost (Savings)	$	1,535
Net Taxes & Fees	$	144
Net Financing	$	405
Net Insurance	$	509
Relative Value Loss	$	-
Net Cost (Savings)	$	5,238

Social Benefits		
Refueling Time Savings	$	15
Energy Security	$	21
Social Cost of Carbon Emissions	$	670
Crashes, Fatalities, Congestion, Noise	$	13
Lost Fuel Tax Revenue	$	(304)
Net Cost (Savings) Including Social Benefits	$	4,824

Meets 2020 Target	✓
Meets Original 2025 Target	✓
Saves Money (including Social Benefits)	✗
Breakeven cost of Gasoline (per Gallon)	$ 3.38

CAFE Fuel Economy (mpg)	35
On Road mpg Winter (EPA 5-cycle Study)	30
On Road mpg Summer (EPA 5-cycle Study)	28
On Road mpg Typical Driving - Accessories OFF (EPA 5-cycle Study)	33

Assumes Diesel and Electricity Prices do not change in relation to gasoline

RAM 1500 4X4

MSRP: $ 35,178
Gasoline Counterpart: 1500 4X4

Lifetime Economic Costs		
Technology Cost	$	2,645
Net Fuel Cost (Savings)	$	3,879
Net Taxes & Fees	$	144
Net Financing	$	405
Net Insurance	$	509
Relative Value Loss	$	-
Net Cost (Savings)	$	7,582

Social Benefits		
Refueling Time Savings	$	13
Energy Security	$	18
Social Cost of Carbon Emissions	$	670
Crashes, Fatalities, Congestion, Noise	$	33
Lost Fuel Tax Revenue	$	(301)
Net Cost (Savings) Including Social Benefits	$	7,149

Meets 2020 Target	✓
Meets Original 2025 Target	✗
Saves Money (including Social Benefits)	✗
Breakeven cost of Gasoline (per Gallon)	$ 3.38

CAFE Fuel Economy (mpg)	32
On Road mpg Winter (EPA 5-cycle Study)	28
On Road mpg Summer (EPA 5-cycle Study)	26
On Road mpg Typical Driving - Accessories OFF (EPA 5-cycle Study)	31

Assumes Diesel and Electricity Prices do not change in relation to gasoline

Diesel

Chevrolet COLORADO 2WD

MSRP: $ 37,625
Gasoline Counterpart: COLORADO 2WD

Lifetime Economic Costs		
Technology Cost	$	9,830
Net Fuel Cost (Savings)	$	4,917
Net Taxes & Fees	$	537
Net Financing	$	1,506
Net Insurance	$	1,890
Relative Value Loss	$	-
Net Cost (Savings)	$	18,680

Social Benefits		
Refueling Time Savings	$	6
Energy Security	$	9
Social Cost of Carbon Emissions	$	503
Crashes, Fatalities, Congestion, Noise	$	41
Lost Fuel Tax Revenue	$	(150)
Net Cost (Savings) Including Social Benefits	$	18,270

Meets 2020 Target	✓	
Meets Original 2025 Target	✗	
Saves Money (including Social Benefits)	✗	
Breakeven cost of Gasoline (per Gallon)	$	5.00

CAFE Fuel Economy (mpg)	31
On Road mpg Winter (EPA 5-cycle Study)	27
On Road mpg Summer (EPA 5-cycle Study)	25
On Road mpg Typical Driving - Accessories OFF (EPA 5-cycle Study)	30

Diesel

Chevrolet COLORADO 4WD

MSRP: $ 41,140
Gasoline Counterpart: COLORADO 4WD

Lifetime Economic Costs	
Technology Cost	$ 8,345
Net Fuel Cost (Savings)	$ 4,987
Net Taxes & Fees	$ 456
Net Financing	$ 1,278
Net Insurance	$ 1,605
Relative Value Loss	$ -
Net Cost (Savings)	$ 16,671

Social Benefits	
Refueling Time Savings	$ 6
Energy Security	$ 9
Social Cost of Carbon Emissions	$ 521
Crashes, Fatalities, Congestion, Noise	$ 42
Lost Fuel Tax Revenue	$ (165)
Net Cost (Savings) Including Social Benefits	$ 16,257

Meets 2020 Target	✓
Meets Original 2025 Target	✗
Saves Money (including Social Benefits)	✗
Breakeven cost of Gasoline (per Gallon)	$ 4.62

CAFE Fuel Economy (mpg)	29
On Road mpg Winter (EPA 5-cycle Study)	25
On Road mpg Summer (EPA 5-cycle Study)	24
On Road mpg Typical Driving - Accessories OFF (EPA 5-cycle Study)	29

Assumes Diesel and Electricity Prices do not change in relation to gasoline

Diesel

Chevrolet COLORADO ZR2 4WD

MSRP: $ 47,595
Gasoline Counterpart: COLORADO ZR2 4WD

Lifetime Economic Costs		
Technology Cost	$	3,500
Net Fuel Cost (Savings)	$	4,883
Net Taxes & Fees	$	191
Net Financing	$	536
Net Insurance	$	673
Relative Value Loss	$	-
Net Cost (Savings)	$	9,784

Social Benefits		
Refueling Time Savings	$	7
Energy Security	$	10
Social Cost of Carbon Emissions	$	617
Crashes, Fatalities, Congestion, Noise	$	41
Lost Fuel Tax Revenue	$	(259)
Net Cost (Savings) Including Social Benefits	$	9,368

Meets 2020 Target	✓
Meets Original 2025 Target	✗
Saves Money (including Social Benefits)	✗
Breakeven cost of Gasoline (per Gallon)	$ 3.52

CAFE Fuel Economy (mpg)	25
On Road mpg Winter (EPA 5-cycle Study)	21
On Road mpg Summer (EPA 5-cycle Study)	21
On Road mpg Typical Driving - Accessories OFF (EPA 5-cycle Study)	26

Assumes Diesel and Electricity Prices do not change in relation to gasoline

Diesel

MSRP: $ 45,600
Gasoline Counterpart: SILVERADO 2WD

Lifetime Economic Costs	
Technology Cost	$ 15,705
Net Fuel Cost (Savings)	$ 3,089
Net Taxes & Fees	$ 857
Net Financing	$ 2,406
Net Insurance	$ 3,020
Relative Value Loss	$ -
Net Cost (Savings)	$ 25,078

Social Benefits	
Refueling Time Savings	$ 18
Energy Security	$ 26
Social Cost of Carbon Emissions	$ 756
Crashes, Fatalities, Congestion, Noise	$ 26
Lost Fuel Tax Revenue	$ (383)
Net Cost (Savings) Including Social Benefits	$ 24,635

Meets 2020 Target	✓
Meets Original 2025 Target	✓
Saves Money (including Social Benefits)	✗
Breakeven cost of Gasoline (per Gallon)	$ 5.97

CAFE Fuel Economy (mpg)	36
On Road mpg Winter (EPA 5-cycle Study)	31
On Road mpg Summer (EPA 5-cycle Study)	28
On Road mpg Typical Driving - Accessories OFF (EPA 5-cycle Study)	34

Assumes Diesel and Electricity Prices do not change in relation to gasoline

Diesel

Chevrolet SILVERADO 4WD

MSRP: $ 49,200
Gasoline Counterpart: SILVERADO 4WD

Lifetime Economic Costs	
Technology Cost	$ 16,060
Net Fuel Cost (Savings)	$ 3,024
Net Taxes & Fees	$ 877
Net Financing	$ 2,460
Net Insurance	$ 3,088
Relative Value Loss	$ -
Net Cost (Savings)	$ 25,510

Social Benefits	
Refueling Time Savings	$ 18
Energy Security	$ 26
Social Cost of Carbon Emissions	$ 805
Crashes, Fatalities, Congestion, Noise	$ 26
Lost Fuel Tax Revenue	$ (423)
Net Cost (Savings) Including Social Benefits	$ 25,059

Meets 2020 Target	✓
Meets Original 2025 Target	✓
Saves Money (including Social Benefits)	✗
Breakeven cost of Gasoline (per Gallon)	$ 5.83

CAFE Fuel Economy (mpg)	33
On Road mpg Winter (EPA 5-cycle Study)	29
On Road mpg Summer (EPA 5-cycle Study)	27
On Road mpg Typical Driving - Accessories OFF (EPA 5-cycle Study)	32

Assumes Diesel and Electricity Prices do not change in relation to gasoline

Diesel

GMC CANYON 2WD

MSRP: $ 38,800
Gasoline Counterpart: CANYON 2WD

Lifetime Economic Costs	
Technology Cost	$ 10,105
Net Fuel Cost (Savings)	$ 4,917
Net Taxes & Fees	$ 552
Net Financing	$ 1,548
Net Insurance	$ 1,943
Relative Value Loss	$ -
Net Cost (Savings)	$ 19,065

Social Benefits	
Refueling Time Savings	$ 6
Energy Security	$ 9
Social Cost of Carbon Emissions	$ 503
Crashes, Fatalities, Congestion, Noise	$ 41
Lost Fuel Tax Revenue	$ (150)
Net Cost (Savings) Including Social Benefits	$ 18,655

Meets 2020 Target	✓
Meets Original 2025 Target	✗
Saves Money (including Social Benefits)	✗
Breakeven cost of Gasoline (per Gallon)	$ 5.06

CAFE Fuel Economy (mpg)	31
On Road mpg Winter (EPA 5-cycle Study)	27
On Road mpg Summer (EPA 5-cycle Study)	25
On Road mpg Typical Driving - Accessories OFF (EPA 5-cycle Study)	30

Assumes Diesel and Electricity Prices do not change in relation to gasoline

Diesel

GMC CANYON 4WD

MSRP: $ 42,340
Gasoline Counterpart: CANYON 4WD

Lifetime Economic Costs	
Technology Cost	$ 5,245
Net Fuel Cost (Savings)	$ 4,987
Net Taxes & Fees	$ 286
Net Financing	$ 804
Net Insurance	$ 1,009
Relative Value Loss	$ -
Net Cost (Savings)	$ 12,330

Social Benefits	
Refueling Time Savings	$ 6
Energy Security	$ 9
Social Cost of Carbon Emissions	$ 521
Crashes, Fatalities, Congestion, Noise	$ 42
Lost Fuel Tax Revenue	$ (165)
Net Cost (Savings) Including Social Benefits	$ 11,917

Meets 2020 Target	✓
Meets Original 2025 Target	✗
Saves Money (including Social Benefits)	✗
Breakeven cost of Gasoline (per Gallon)	$ 4.02

CAFE Fuel Economy (mpg)	29
On Road mpg Winter (EPA 5-cycle Study)	25
On Road mpg Summer (EPA 5-cycle Study)	24
On Road mpg Typical Driving - Accessories OFF (EPA 5-cycle Study)	29

Assumes Diesel and Electricity Prices do not change in relation to gasoline

Diesel

GMC SIERRA 2WD

MSRP: $ 51,760
Gasoline Counterpart: SIERRA 2WD

Lifetime Economic Costs	
Technology Cost	$ 10,165
Net Fuel Cost (Savings)	$ 3,433
Net Taxes & Fees	$ 555
Net Financing	$ 1,557
Net Insurance	$ 1,955
Relative Value Loss	$ -
Net Cost (Savings)	$ 17,665

Social Benefits	
Refueling Time Savings	$ 16
Energy Security	$ 22
Social Cost of Carbon Emissions	$ 707
Crashes, Fatalities, Congestion, Noise	$ 29
Lost Fuel Tax Revenue	$ (338)
Net Cost (Savings) Including Social Benefits	$ 17,228

Meets 2020 Target	✓
Meets Original 2025 Target	✓
Saves Money (including Social Benefits)	✗
Breakeven cost of Gasoline (per Gallon)	$ 4.88

CAFE Fuel Economy (mpg)	35
On Road mpg Winter (EPA 5-cycle Study)	30
On Road mpg Summer (EPA 5-cycle Study)	28
On Road mpg Typical Driving - Accessories OFF (EPA 5-cycle Study)	33

Assumes Diesel and Electricity Prices do not change in relation to gasoline

Diesel

GMC SIERRA 4WD

MSRP: $ 54,560
Gasoline Counterpart: SIERRA 4WD

Lifetime Economic Costs	
Technology Cost	$ 10,165
Net Fuel Cost (Savings)	$ 8,638
Net Taxes & Fees	$ 555
Net Financing	$ 1,557
Net Insurance	$ 1,955
Relative Value Loss	$ -
Net Cost (Savings)	$ 22,870

Social Benefits	
Refueling Time Savings	$ (10)
Energy Security	$ (15)
Social Cost of Carbon Emissions	$ (38)
Crashes, Fatalities, Congestion, Noise	$ 73
Lost Fuel Tax Revenue	$ 322
Net Cost (Savings) Including Social Benefits	$ 22,538

Meets 2020 Target	✓
Meets Original 2025 Target	✗
Saves Money (including Social Benefits)	✗
Breakeven cost of Gasoline (per Gallon)	$ 5.50

CAFE Fuel Economy (mpg)	24
On Road mpg Winter (EPA 5-cycle Study)	21
On Road mpg Summer (EPA 5-cycle Study)	20
On Road mpg Typical Driving - Accessories OFF (EPA 5-cycle Study)	25

Assumes Diesel and Electricity Prices do not change in relation to gasoline

Diesel

GMC SIERRA 4WD AT4

MSRP: $ 69,185
Gasoline Counterpart: SIERRA 4WD AT4

Lifetime Economic Costs		
Technology Cost	$	9,890
Net Fuel Cost (Savings)	$	6,841
Net Taxes & Fees	$	540
Net Financing	$	1,515
Net Insurance	$	1,902
Relative Value Loss	$	-
Net Cost (Savings)	$	20,688

Social Benefits		
Refueling Time Savings	$	(1)
Energy Security	$	(1)
Social Cost of Carbon Emissions	$	287
Crashes, Fatalities, Congestion, Noise	$	58
Lost Fuel Tax Revenue	$	31
Net Cost (Savings) Including Social Benefits	$	20,314

Meets 2020 Target	✓	
Meets Original 2025 Target	✗	
Saves Money (including Social Benefits)	✗	
Breakeven cost of Gasoline (per Gallon)	$	4.94

CAFE Fuel Economy (mpg)		24
On Road mpg Winter (EPA 5-cycle Study)		21
On Road mpg Summer (EPA 5-cycle Study)		20
On Road mpg Typical Driving - Accessories OFF (EPA 5-cycle Study)		25

Assumes Diesel and Electricity Prices do not change in relation to gasoline

Land Rover Discovery

MSRP: $ 54,300
Gasoline Counterpart: Discovery

Lifetime Economic Costs	
Technology Cost	$ 2,000
Net Fuel Cost (Savings)	$ 2,066
Net Taxes & Fees	$ 109
Net Financing	$ 306
Net Insurance	$ 385
Relative Value Loss	$ -
Net Cost (Savings)	$ 4,866

Social Benefits	
Refueling Time Savings	$ 21
Energy Security	$ 29
Social Cost of Carbon Emissions	$ 1,033
Crashes, Fatalities, Congestion, Noise	$ 17
Lost Fuel Tax Revenue	$ (623)
Net Cost (Savings) Including Social Benefits	$ 4,389

Meets 2020 Target	✓
Meets Original 2025 Target	✗
Saves Money (including Social Benefits)	✗
Breakeven cost of Gasoline (per Gallon)	$ 2.94

CAFE Fuel Economy (mpg)	30
On Road mpg Winter (EPA 5-cycle Study)	26
On Road mpg Summer (EPA 5-cycle Study)	25
On Road mpg Typical Driving - Accessories OFF (EPA 5-cycle Study)	30

Assumes Diesel and Electricity Prices do not change in relation to gasoline

Land Rover Range Rover

MSRP: $ 91,700
Gasoline Counterpart: Range Rover

Lifetime Economic Costs	
Technology Cost	$ 800
Net Fuel Cost (Savings)	$ 1,242
Net Taxes & Fees	$ 44
Net Financing	$ 123
Net Insurance	$ 154
Relative Value Loss	$ -
Net Cost (Savings)	$ 2,362

Social Benefits	
Refueling Time Savings	$ 26
Energy Security	$ 37
Social Cost of Carbon Emissions	$ 1,143
Crashes, Fatalities, Congestion, Noise	$ 10
Lost Fuel Tax Revenue	$ (723)
Net Cost (Savings) Including Social Benefits	$ 1,869

Meets 2020 Target	✓
Meets Original 2025 Target	✗
Saves Money (including Social Benefits)	✗
Breakeven cost of Gasoline (per Gallon)	$ 2.64

CAFE Fuel Economy (mpg)	32
On Road mpg Winter (EPA 5-cycle Study)	28
On Road mpg Summer (EPA 5-cycle Study)	26
On Road mpg Typical Driving - Accessories OFF (EPA 5-cycle Study)	31

Assumes Diesel and Electricity Prices do not change in relation to gasoline

Land Rover Range Rover Sport

MSRP: $ -
Gasoline Counterpart: Range Rover Sport

Lifetime Economic Costs		
Technology Cost	$	-
Net Fuel Cost (Savings)	$	1,792
Net Taxes & Fees	$	-
Net Financing	$	-
Net Insurance	$	-
Relative Value Loss	$	-
Net Cost (Savings)	$	1,792

Social Benefits		
Refueling Time Savings	$	24
Energy Security	$	34
Social Cost of Carbon Emissions	$	1,046
Crashes, Fatalities, Congestion, Noise	$	15
Lost Fuel Tax Revenue	$	(637)
Net Cost (Savings) Including Social Benefits	$	1,310

Meets 2020 Target	✓	
Meets Original 2025 Target	✗	
Saves Money (including Social Benefits)	✗	
Breakeven cost of Gasoline (per Gallon)	$	2.56

CAFE Fuel Economy (mpg)	32
On Road mpg Winter (EPA 5-cycle Study)	28
On Road mpg Summer (EPA 5-cycle Study)	26
On Road mpg Typical Driving - Accessories OFF (EPA 5-cycle Study)	31

Assumes Diesel and Electricity Prices do not change in relation to gasoline

OTHER ADVANCED TECHNOLOGIES

The 2020 Model Year EPA/NHTSA Fuel Economy Guide contains information on six hundred eighty-one vehicles with **eight or more forward gears. Twenty percent of these vehicles meet their 2020 model year fuel economy targets** and six percent meet their 2025 model year fuel economy targets.

The 2020 Model Year EPA/NHTSA Fuel Economy Guide contains information on five hundred ninety vehicles with **start-stop technology. Twenty-one percent of these vehicles meet their 2020 model year fuel economy targets** and six percent meet their 2025 model year fuel economy targets.

The 2020 Model Year EPA/NHTSA Fuel Economy Guide contains information on six hundred fifteen vehicles with **turbochargers or superchargers. Fifteen percent of these vehicles meet their 2020 model year fuel economy targets** and one percent meet their 2025 model year fuel economy targets.

One hundred eighty-three vehicles in the 2020 model year database contain the key technologies EPA claims are required to meet the 2025 model year fuel economy target including variable valve timing, variable valve lift, eight or more forward gears, and start stop technology. This category excludes electric vehicles, plug-in hybrids, hybrid vehicles and diesel vehicles which EPA claims is not required technology. **Of the vehicles that contain all of these technologies, fifteen percent met their 2020 model year fuel economy target and less than two percent met their 2025 model year fuel economy target.**

The 2020 Model Year EPA/NHTSA Fuel Economy Guide contains information on **eleven hundred and nineteen vehicles. Twenty-two percent of these vehicles met**

their 2020 model year fuel economy targets and ten percent met their 2025 model year fuel economy targets.

Of the one hundred and nine vehicles that met their 2025 model year target twenty-six are all electric, nineteen are plug-in hybrids, twenty-eight are hybrid electric vehicles, and twelve are powered by diesel engines. The average net cost[5] of the technology to comply with the 2025 model year fuel economy targets is $11,800 above the cost of the average 2020 model year vehicle.

---

[5] Technology cost minus fuel savings minus social benefits

METHODOLOGY

The VOLPE CAFE model has been a staple in the regulatory arena for well over a decade and has gone through a number of iterations during that time.

The methodology used in this analysis is derived from the June 2018 updated version of the VOLPE CAFE model.

The cost of the technology was based on the difference in the manufacturer's suggested retail price (MSRP) as listed on the manufacturer's websites for the advanced technology vehicle compared to the MSRP for the base counterpart gasoline vehicle without the advanced technology.

The fuel economy benefit was based on the difference in the EPA measured fuel economy (unadjusted combined city – highway) for the vehicle with the advanced technology compared to the base vehicle without the advanced technology.

The on-road "gap" used in the VOLPE analysis to adjust the tested fuel economy values relies on a static gap of twenty percent. Since 2008, EPA has been using a dynamic "gap" to adjust test fuel economy values for fuel economy labeling purposes to better reflect "on-road" fuel economy. This analysis uses the EPA dynamic "gap" for gasoline vehicles and a separate dynamic "gap" for hybrid electric vehicles developed using the EPA methodology.

Fuel economy data used in this study is taken from the EPA website 2020 Fuel Economy Guide for DOE with release dates before January 3, 2020.

SUMMARY

2017 Model Year

Advanced Technology	BEV	PHEV	HEV	Diesel	8 or More Gears	Start/Stop	Turbo or Supercharged	All
Number Evaluated	11	14	43	16	435	386	631	1248
Number Less Expensive to Own	0	1	3	4				
Percent that Save Customer Money Over Lifetime	0%	7%	7%	25%				
Met 2017 Target	100%	100%	91%	100%	20%	29%	17%	23%
Met 2025 Target	100%	86%	70%	50%	3%	21%	0%	4%

2018 Model Year

Advanced Technology	BEV	PHEV	HEV	Diesel	8 or More Gears	Start/Stop	Turbo or Supercharged	All
Number Evaluated	10	25	33	23	467	529	648	1209
Number Less Expensive to Own	0	0	1	0				
Percent that Save Customer Money Over Lifetime	0%	0%	3%	0%				
Met 2018 Target	100%	96%	91%	96%	17%	24%	21%	28%
Met 2025 Target	100%	68%	58%	22%	0.2%	5%	0.8%	4%

2019 Model Year

Advanced Technology	BEV	PHEV	HEV	Diesel	8 or More Gears	Start/Stop	Turbo or Supercharged	Key Technologies	All
Number Evaluated	24	19	35	25	616	617	688	123	1240
Number Less Expensive to Own	0	0	2	2					
Percent that Save Customer Money Over Lifetime	0%	0%	6%	8%					
Met 2019 Target	100%	100%	97%	92%	12%	11%	10%	17%	23%
Met 2025 Target	100%	84%	71%	28%	1.0%	3%	1.0%	0.3%	6%
Number meeting 2025 Target	24	16	25	7	6	19	7	0	77

2020 Model Year

Advanced Technology	BEV	PHEV	HEV	Diesel	8 or More Gears	Start/Stop	Turbo or Supercharged	Key Technologies	All	All Non BEV/PHEV
Number Evaluated	26	30	72	16	681	590	615	183	1119	1063
Number Less Expensive to Own	0	0	8	0						
Percent that Save Customer Money Over Lifetime	0%	0%	11%	0%						
Met 2020 Target	100%	80%	51%	69%	20%	21%	15%	15%	22%	19%
Met 2025 Target	100%	63%	39%	25%	6.0%	5.6%	0.8%	1.6%	10%	6%
Number meeting 2020 Target	26	24	37	11	138	125	91	28	248	198
Number meeting 2025 Target	26	19	29	4	17	33	5	3	109	64

ABOUT THE AUTHOR

Walter Kreucher spent the better part of his thirty-year career in the automobile industry working on fuel economy and mobile source issues. He helped Ford institute the very first preliminary CAFÉ compliance procedures over forty years ago and later in his career ran the CAFE compliance activity. After retirement he consulted
for NHTSA and VOLPE helping them address critical issues within the context of the CAFÉ program. He has also consulted with environmental organizations that worked "the other side of the aisle".

www.ingramcontent.com/pod-product-compliance
Lightning Source LLC
Chambersburg PA
CBHW052359220526
45465CB00003BB/1180